Building Maintenance

Brian Wood, DipArch MCMI RIBA MRICS MBIFM FCIOB

Department of Real Estate and Construction
School of the Built Environment
Oxford Brookes University

A John Wiley & Sons, Ltd., Publication

This edition first published 2009
© 2009 by Blackwell Publishing Ltd

Blackwell Publishing was acquired by John Wiley & Sons in February 2007. Blackwell's publishing programme has been merged with Wiley's global Scientific, Technical, and Medical business to form Wiley-Blackwell.

Registered office
John Wiley & Sons Ltd, The Atrium, Southern Gate, Chichester, West Sussex, PO19 8SQ, United Kingdom

Editorial offices
9600 Garsington Road, Oxford, OX4 2DQ, United Kingdom
2121 State Avenue, Ames, Iowa 50014-8300, USA

For details of our global editorial offices, for customer services and for information about how to apply for permission to reuse the copyright material in this book please see our website at www.wiley.com/wiley-blackwell.

Library of Congress Cataloging-in-Publication Data

Wood, Brian.
   Building maintenance/Brian Wood.
      p. cm.
   Includes bibliographical references and index.
   ISBN 978-1-4051-7967-6 (pbk. : alk. paper)   1. Buildings–Maintenance.   I. Title.

   TH3351.W663 2009
   690′.24–dc22

                                                                2009001746

A catalogue record for this book is available from the British Library.

Set in 9.5/11.5 pt Minion by Aptara® Inc., New Delhi, India
Printed in Singapore

1   2009

Contents

Foreword

In his Foreword to *Building Care*, Sir Michael Latham (author of the seminal *Constructing the Team*) said that Brian Wood rightly confronts the dismissive attitude to building maintenance throughout his thoughtful and profound analysis of the maintenance process.

In this book Brian continues to apply that incisive mind to questioning existing maintenance practice so as to present here an instructive and comprehensive examination of maintenance processes and to promulgate and promote good practice. He has been working in this much-neglected field now for over 30 years, building a vast mine of information and experience that he shares with you here. It is often said that maintenance is not 'sexy'. Yet it is 'big bucks'. Much more will be spent on maintaining buildings than on constructing them in the first place, so it is worth trying to learn from the past and to build better buildings.

Building maintenance and facilities managers must be at the centre of deliberations on the briefing and design of new buildings as well as the refurbishment, rearrangement and repair of existing ones. They should have a seat at the boardroom table, not be brought in once all the decisions have been made and be expected to somehow make everything work. After staff costs, buildings are the next highest – they also offer great scope for improving staff morale, well-being and productivity. They also of course need to be run efficiently, effectively and economically.

Sustainability is key. No serious attempt to save the planet by making new buildings 'zero carbon' can have enough impact, other than in terms of showing what is possible. The focus has to be on the buildings that exist already. Every repair action on a building element has within it the potential to improve performance of the building as a whole; these opportunities must be seen and seized.

An increasingly professional approach is required, with decisions based on evidence and experience. A strategic approach is required if buildings are going to be operated to best effect. Maintenance and facilities managers will be encouraged to stand back to gain perspective and to consider a range of alternative futures. The use of checklists and case studies will also enable them to learn from and apply experience to their day-to-day tasks.

The author has a passion for improving the image and practice of building maintenance, and this shines through throughout the book. I hope the book will have a large audience and help its readers share that passion and thereby make a positive difference through their work too.

Professor Jim Smith
Head of the School of Sustainable Development
Bond University
Gold Coast, Queensland
Australia

Preface

This book is intended to help building maintenance and facilities managers to do their jobs better. Unfortunately, few practitioners feel the need or find the time to read books to keep up to date with new ideas and knowledge of what others are doing and thereby to improve their own practice. Everyone has busy lives. So I have aimed the book at an audience that I hope will include students – the practitioners of tomorrow.

Building maintenance is expensive; it costs many times more to run the building over its lifetime than to build it, yet maintenance is not accorded the priority it warrants; nor are maintenance practitioners given the respect, let alone remuneration, that they deserve to have. A poorly maintained building will be a drain on increasingly limited resources, expensive to run and potentially dangerous. By contrast, a well maintained building is an appreciating asset, an investment in the future and in those who work in or use the building.

Maintenance and facilities managers should have input at the design stage of new buildings and in the preparation of improvement plans; they will have responsibility for the continuing care of the assets. They should be able to make an important and worthwhile contribution to the development of strategy, policies and procedures. At the same time they need to be responsive and to deal with problems as they arise. The book addresses both long-term and day-to-day issues, providing practical ways forward.

People are at the core – the client, building users, the maintenance team, visitors and passers-by; all will have views on what should be done, when, how and who by. The maintenance manager needs not only the knowledge to determine the 'right' or best technical solution, but also the group-working and interpersonal skills to make things happen on the ground. Our buildings exist for people; we can but try to help them work well. The book is aimed at helping maintenance managers have easier and more rewarding lives by reducing ongoing demands and making a positive difference.

Brian Wood

Acknowledgements

This book would not have happened without a lot of encouragement. We must encourage each other, especially when we are all under pressure – to produce results, outputs and solutions to problems. There is no shortage of things to do, including many that we are told must be done. The pressure means that sometimes we (sorry, I mean I!) get irritated and short-tempered or 'blow up'; and take it out on those trying to help, who we then fail to even think to thank. So I want to thank many people, and I crave forgiveness from those I neglect to thank here.

I owe a great debt to my wife Erica in helping me get it together and in improving the work by questioning why and how I say what I do: the model of a 'critical friend'. I have had many critics through my career and I would like to thank those who have helped me thereby. I therefore thank my colleagues and students at Oxford Brookes University, where I have been teaching and developing my thoughts since 1990, for confirming the value of what I have to say.

I have been privileged to have had the opportunity also to share my thoughts with audiences around the world, from Aalborg and Auckland to Watford and Weston-super-Mare. I should, I guess, be sad about all the air-miles incurred but I am so glad to have been able to make such a great number of great friendships thereby and to enjoy wonderful hospitality. I want to thank particularly Bob Branch, Eddie Finch, John Hornibrook, Lo Sui Pheng, Geoff Outhred, Marie-Cecile Puybaraud, Hedley Smyth, Rosemary Schofield and Suzanne Wilkinson, each of whom thought I had something worthwhile to return in friendship.

Finally I must acknowledge my genetic inheritance. Although my father died some 30 years ago I continue to benefit from his determination and perseverance, and that I see continuing in my son Tom, who married this year and has everything ahead. I have had sadness this year too though – my mother died. She had been unwell since having a stroke, but was able to spend her last year well looked after in a care home where staff from all round the world happily shared their time and love with my mum. I hope to be able to share some of her supportive non-judgementalism and acceptance of suboptimal performance and 'failure' when marking student's assignments.

Thanks to all who have helped in whatever way to create this book.

1 Introduction

Building maintenance will always be with us. Although the maintenance-free building may be a theoretical (and increasingly technical) possibility, all buildings are subject to the vagaries of weather, decay and variations in use that necessitate changes and updating of the building and its component parts. However, it is often neglected. Maintenance is not sexy. It has been described as the 'Cinderella' of the building industry (Seeley, 1976). Cinderella, however, did get to go to the ball.

A lot has changed since Seeley used the Cinderella appellation in 1976. Wordsworth (2001) and Chanter and Swallow (2007) have remarked upon the rate of change in management interest in this field. With the growth in the profession and practice of facilities management in the late twentieth century and a belatedly growing interest in sustainability, whoever is responsible for the design, construction and management of buildings is at the centre of decisions vital to economic success, well-being and saving the planet.

In my earlier book, *Building Care*, I suggested that a significant rethinking of building maintenance was required. I still believe that. But the construction industry generally, and those involved with existing buildings more particularly, are fundamentally quite conservative rather than radical. Perhaps this is to be expected when 'conserving' buildings. This book is targeted specifically at enabling all those involved with building maintenance, or preparing to be, to do the job required from day to day.

Building Care applies a people-centred approach to building maintenance; it puts building clients and occupants centre-stage in determining priorities across the board. That remains important, but is not at the core of this book, which focuses on the work of the maintenance manager and his or her team. There is a strong practical aspect to this book.

Arrangement of this book

The opening chapters review the context in which building maintenance is carried out and the vital contribution that a good building design can make to good maintenance. The chapters following are aimed at helping the maintenance team organise the work effectively through planning and prioritisation. There is a substantial section (totalling six chapters) dedicated to dealing with defects – their identification and rectification – in more detail. The book concludes with chapters that look at the bigger picture, upgrading and regenerating buildings, bringing new life to them and the environments in which they are situated.

Problems

Much of the maintenance manager's work, and that of operatives too for that matter, centres on problem solving. It would be good to give more attention to problem *avoidance* (and attention is given to this aspect in Chapter 8), but we must still deal with the problem when it is presented to us.

Problem (noun):

- Doubtful or difficult question.
- Thing hard to understand.
- Proposition in which something has to be done.
- The question.
- Inquiry starting from given conditions to investigate a fact, result or law.
- A challenge to accomplish a specified result, often under prescribed conditions.

(Fowler *et al.*, 1934)

- A puzzle; a riddle; an enigmatic statement.
- A difficult or demanding question; a matter or situation regarded as unwelcome, harmful or wrong, and needing to be overcome; a difficulty.
- The quality of being problematic or difficult to solve.

(Oxford English Dictionary Online, 2008)

When some part of the building or its equipment is broken or has failed, something must be done, and hopefully soon (preferably yesterday!). The book has a problem-solving emphasis. Typical problems are introduced and discussed with a view to their solution.

The word 'problem' can be understood, and often is, in a rather negative way; it suggests that something has gone wrong that ought not to have, and that by inference it is someone's fault, someone is to blame and they ought to put it right and/or pay. Sometimes this can be unhelpful with the focus becoming one of blaming, refutation and counter-blaming, sapping time and energy that would be better applied to problem solving and implementation. Perhaps the word 'issue' would be perceived as less negative.

The construction industry has a strongly developed adversarial culture. This has been commented upon in official reports for over half a century (Simon, 1944; Emmerson, 1962; Banwell, 1964; National Economic Development Office (NEDO), 1964, 1983; Wood, 1975; Latham, 1994; Egan, 1998; Fairclough, 2002). This culture is generated and fuelled by the competitive arrangements for procuring buildings. Competitive tendering encourages a focus on cheapness and hoping to 'cut corners' to get the job done for the price quoted. Time will be spent trying to find ways of claiming for 'extras' and/or related to time lost due to changing instructions, unavailability of materials and hold-ups due to others, etc. This is time that could be put to the better execution of the work if harnessed constructively. There is more detail on this in relation to maintenance contracts in Chapter 9.

The maintenance team will often, though not always, be able to draw upon a range of expertise and experience, if they are not at each others' throats. It is thus important for those in the maintenance team to be able to get on with others, to seek and to respect their views, as well as to have a view to the best technical solution.

Checklists

Several checklists are developed in this book to assist in the observation of problems and the determination of appropriate solutions.

A checklist has much to commend it. It is normally clearly set out, based on previous experience, and easy and quick to complete. It is important, however, to keep in mind the limitations of checklists – their strengths can also be weaknesses; they can be rather reductionist. Checklists can be either too short or too detailed. Short checklists have advantages in terms of speed and simplicity of application, but they may miss some important aspect for the particular situation. A long list may be thought to be comprehensive but it may still be missing that particular and unexpected something. There can be great value in anticipating this possibility by adding a heading or two labelled 'other' to prompt the surveyor to think and look more widely.

Problems generally present themselves through symptoms or manifestations. Rather as a headache only tells you directly that your head hurts, it doesn't tell you the cause. The headache may be a symptom of a bang on the head, workplace stress, a brain tumour or a stroke, or any one of a range of possibilities not considered. In the same way with a building, an ingress of water at a window head doesn't tell us whether there is a defective lintel, porous brickwork, or a leaking gutter or roof above. It may stem from a leaking hot or cold water pipe or radiator in the attic or from a floor above. The maintenance manager needs the skills of a pathologist or forensic scientist to consider the possibilities and to 'follow the trail' of symptoms back to the source. A checklist may help here; or it may inhibit the consideration of more unusual or unlikely causes.

A particular virtue of the checklist is that it forces the consideration of the issues listed and, if so designed, requires the surveyor to make a mark of some kind against it, thus offering an audit trail, demonstrating that the particular issue was considered. This is particularly useful where a surveyor may be inspecting a particular building or building type with which they are unfamiliar. It is also useful where a team of surveyors is involved and it is necessary to ensure consistency of standards of assessment. It provides a record of the survey; it also enables tracking of work done and comparison of condition over time if a consistent checklist is used.

Maintenance: what is it; what is it for?

To start with some basics, it is important to set out terms of reference, including some definitions. If people are not agreed on what maintenance is, and what is intended through maintenance, then time and effort are likely to be used ineffectively. Of course there will always be plenty of scope for disagreement on the best way of proceeding on any matter, but it is good to agree on the desired end. It is also important that when a client or their agent, perhaps a surveyor, is ordering a particular action from an operative or contractor that there is a clear and common understanding of what is expected.

Definitions

Definitions evolve and change over time as understanding develops and as the context in which they are applied changes. In the UK, the British Standards Institution (BSI) has a long history of involvement in determining appropriate standards related to buildings and construction. BSI was formed as the Engineering Standards Committee in 1901 (www.bsi-global.com). The development and revision of British Standards is informed by committees of advisors drawn from the construction industry, the professions and representatives of central and local government. British Standards represent a compromise between the best that could be conceived and those commonly achieved. They commonly contain clauses controlling the specifications of material and workmanship and set out methods for their testing.

In 1972, the Committee on Building Maintenance commended this definition of maintenance:

> ... *work carried out in order to keep, restore or improve every facility, i.e. every part of the building, its services and surrounds to a currently acceptable standard and to sustain the utility and value of the facility.* (Department of the Environment, 1972)

BS 3811: 1984 (BSI, 1984) adopted this definition of maintenance:

> *A combination of any actions carried out to retain an item in, or restore it to an acceptable condition.*

This definition retained the dual possibility of *keeping* an item at a certain level or of *restoring* it to that position of acceptability, although it does not identify who it is that determines the acceptability. The definition drops the references to *every facility* and to *utility* and *value* and to the *services and surrounds* of the building; it substitutes *acceptable standard* with *acceptable condition*. The emphasis is thus shifted from how well a facility operates to the condition of individual items, and thereby from the client or building user interested in value and utility to the surveyor trained in the technical assessment of condition.

This was rebalanced somewhat in BS 8210: 1986 (BSI, 1986), which defined building maintenance as:

> *Work, other than daily and routine cleaning, necessary to maintain the performance of building fabric and its services.*

The differences can be clearly seen. *Daily and routine cleaning* are specifically excluded, but not non-routine cleaning, and only *necessary* work is included; work related to the building's *service*s are again specifically included. The standard to be applied relates to the *performance* of the building and its services, not their *condition*. This might have been expected to encourage a more objective and quantitative determination of performance as opposed to a more qualitative, subjective assessment of condition.

The revision of BS 3811 in 1993 (BSI, 1993) refers to BS 4778, Part 3, Section 3.1 (BSI, 1991) for its definition of maintenance:

> *The combination of all technical and administrative actions, including supervision actions, intended to retain an item in, or restore it to, a state in which it can perform a required function.*

In this definition there is stress on the organisational aspects – administering and supervising – which reflects an increasing attention to and valuing of the management function, although not yet recognising the role of strategy, unless that is to be understood as an administrative task. *Performance* is still preferred to *condition*, but the measure is that the item in question *can* perform the required function, not that it necessarily *does*, and again it is not defined who determines the *required function*.

In addition to these definitions of building maintenance, there are also definitions to consider of various components of maintenance (e.g. maintenance planning, service level agreements) and of various types (e.g. planned maintenance, condition-based maintenance). These will be explored further in Chapter 3.

Shortcomings of definitions

Despite the changes in definition of building maintenance noted and discussed above, the fundamental standard understanding has not moved on very much in a third of a century, despite the repackaging of maintenance within facilities management and the growth of interest in sustainability. At a conference of the Royal Institution of Chartered Surveyors in 2006, I proposed a shorter definition:

> *Building maintenance is the totality of all actions that keep a building functioning effectively.*

Features of this definition are:

- It is more inclusive than the British Standards definitions hitherto.
- It excludes actions that were intended to keep the building in a particular condition or to restore it thereto, but failed so to do.
- It relates to actual performance achieved, not that intended.
- The goal is effectiveness, which is a rolling target; it develops and changes as the building users' needs change.

A problem with definitions is that, as the word suggests, they can be limiting; they tend to 'draw a line' around what is included, and by inference therefore what is not. For instance, the references to retaining or restoring an item to a particular state or condition may cause the maintenance manager's mind to focus on the immediate task and the individual item. Someone needs to be able to stand back, to take the longer view and see the bigger picture. I contend that this too falls within the purview of the maintenance manager. There is limited value in carrying out activities that will restore an item to its original condition if the building is to be radically remodelled or disposed of or demolished soon. I therefore take the view, and follow it in this book, that maintenance includes the whole repair–maintenance–improvement (RMI) continuum.

We may take the view that it is not important outside of the academic environment to have a universally agreed definition of maintenance. What is important is that the right work gets done; but what is the 'right' work – who decides, and on what basis? These issues will be discussed further within this book. Figure 1.1 shows possible components of 'right' building maintenance.

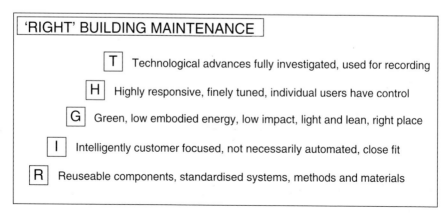

Figure 1.1 Possible components of 'right' building maintenance.

One aspect of definition, whichever definition is chosen, is that it will tend to limit consideration of some aspects. The word 'define' itself suggests a limiting, defining what is in and what is out, included/excluded, acceptable/not acceptable. Another reductionist result of applying definitions stems from the adversarial nature of the construction industry, the need to be able to attach blame and therefore financial liability for anything that didn't meet someone's expectation. Thus attempts will be made to try to define such expectations within a contract, an agreement capable of being enforced in the courts. Much effort and expensive time may be expended in drafting such contracts and in arguing about their interpretation and application. I have heard it suggested that a contract once signed is best left in the drawer or filing cabinet; this is an admirable principle, recognising the value of harmonious working arrangements. However it stands the risk of one party being taken advantage of if the other party is more worldly-wise and better versed in the ways of the courts. It is expensive to go to court and the other party's pockets may be deeper; it can be cheaper to settle without going to court however much one may be 'in the right', and especially if seeking a harmonious relationship from which good references and/or repeat business may stem.

One hope of avoiding such adversarial argument is through the development of partnering (Bennett & Jayes, 1995) and longer-term contracts. The Latham Report (1994) sought more of a team approach to work in the construction industry, building up longer-lasting relationships and more consistent supply chains. The Private Finance Initiative (PFI), in which contractors are responsible for design, construction and maintenance of buildings for maybe 15 or 25 years, allows the possibility of a much greater interest in forecasting and budgeting for the costs of maintenance at the bid stage. PFI and its later incarnation, public–private partnership, thus offer the opportunity of a much greater interest in the longer-term future of buildings and attention to the issue of sustainability.

Sustainability came to the fore as an issue in the 1980s; it is said to be achieved when resources are used in such a way that the quality of life of future generations is ensured (World Commission on Environment and Development (WCED), 1989). That is the purpose and outcome of building maintenance; anything more or less than that is to squander resources. It is also to construe maintenance as a people-centred activity; it is carried out to enable people to function effectively in buildings. There is more on this later; suffice it to say here that while sustainability is very much 'future-focused', building maintenance as generally discussed is dealing today with our inheritance of buildings bequeathed to us by past generations. This book takes it as read that building maintenance should be concerned with yesterday, today and tomorrow. We need to understand the past and present in order to project and plan the future.

Purposes of building maintenance

Buildings exist (and are created) largely for the benefit of their occupants and for what goes on in the building. For instance, an office building exists to facilitate administrative functions; a block of flats is for people to live in, including perhaps some social interactions between residents. The building elements (floors, walls, roofs, etc.) exist largely to divide the building's uses one from another and to keep at bay the external elements of wind, rain, snow, unacceptably high or low temperatures, etc. Building maintenance is carried out in order to allow those functions to continue to be carried out, preferably in the way and to the standard of that originally envisaged, designed and built, and at least satisfactorily.

The layout of the building and its upkeep also have a strong influence on the productivity of its occupants and their well-being (Clements-Croome, 2000). At the same time the asset value of the building may also be affected by its maintenance, or by the lack of it. Building owners may be prepared to accept a declining performance as the building deteriorates over time, calculating perhaps that the cost of maintenance may be greater than the loss in asset value. They may be less prepared to accept declining performance if the building's rental income is reduced, if lessees decline to renew and/or the building becomes unlettable or unsaleable in a declining market.

In Britain there is a strong interest in, and love of, old buildings – whether or not they are listed; it was not always thus, however. The stone buildings of Oxford, so venerated now, were sadly neglected in the 1950s and 1960s and only rescued and restored to their present glory by a group of *aficionados* who set up the Oxford Preservation Trust (Oakeshott, 1975). This growth of respect for old buildings and towns was manifested in books of the day (e.g. Cullen, 1961; Johns, 1965; Worskett, 1969) and given official recognition in the Civic Amenities Act 1967, which created Conservation Areas. In Australia, 'heritage buildings' are similarly valued. The legislation related to these buildings places constraints on the ways they may be altered or added to. While this may be a serious inhibitor of bringing a building, and how it is used, into line with current thinking and fashion, it must be hoped that the heritage value will outweigh the transitory and ephemeral gains accruing from chasing after modernity. There is 'cultural capital' in existing buildings, both individually and perhaps more so in conjunction, in the creation and maintenance of *genius loci* (spirit of the place), the receipt of heritage, its development and its passing on to the next generation and regeneration. No building or space is static. Maintenance is needed whether to keep buildings as they are or to restore them to some previous state, or to bring them to a state in which they meet a new need.

Context of building maintenance

The context in which the particular problem is situated may be significant. A number of case studies are included where there is a degree of complexity and inter-relationship that may influence the appropriateness of a particular choice of solution. For instance, when thinking about repairing or replacing a roof covering it makes sense to consider the condition of a chimney (and what to do about it); it is also sensible to consider the condition of the underlying roof structure. There may be plans for substantial changes to the building in the future that may colour decisions on what action to take, including the possibility of 'living with' a defect for a while. Funds may not be available to undertake all the maintenance work that a surveyor may regard as prudent for the most effective and economical maintenance and operation of the building.

The context may be framed in any number of ways; this section considers five categories of context:

- political;
- economic;
- social;
- technological;
- environmental.

The first four of these are often taken together as what is known as a PEST analysis; it is one way of breaking down a problem into smaller component parts. PEST analysis was developed before environmental (or ecological) issues were accorded the proper consideration that they so clearly warrant today. Further developments and variants include STEP, STEEP, PESTLE, PESTELI, STEEPLE and STEEPLED:

- social;
- technological;
- economic;
- environmental;
- political;
- legal;
- ethical;
- demographic.

The basic principle is the same; it is a methodology for helping to ensure that the individual or team considers a wide (or wider) range of issues related to the problem. Some issues will be placed under one heading in one scheme and under another in another. I am sticking here with PEST, with the addition of 'environmental'. This analysis is not intended to be exhaustive, nor will it be applicable to every building maintenance problem in every location or on each occasion; the contexts of time and place will be important.

Political

This could include aspects such as:

- Government policies and programmes; 'stop–go'/regulator.
- Legal context: laws, regulations; employment; health and safety.
- Skills availability: training, labour mobility and migration.
- Internal politics: who tends to get their way/how do you get yours?

Building maintenance activity is very much affected by government. New and improved buildings and infrastructure are a very tangible sign of economic prosperity. In Britain there is a tendency to associate some political parties more with growth of investment in the public sector, and with others rather less intervention by the state. Over the years since World War II there has been nationalisation, de-nationalisation and re-nationalisation, privatisation and a growth in the voluntary or 'third sector'. Council housing has given way to housing associations and Registered Social Landlords. Large homogeneous council estates have been broken down by sales through the right-to-buy legislation of the conservative government of the 1980s under the then Prime Minister, Mrs (now Baroness) Thatcher. This has effects on the maintenance of those houses. Where previously it may have made sense to develop large programmes of planned maintenance and upgrading, the patchwork of ownership and changing aspirations on the part of occupants suggests a need for a more individualised response.

There has also been substantial investment in new hospitals and other healthcare buildings, and in the Building Schools for the Future (BSF) programme, where every school in England is to be replaced over the coming years (programme subject to future government spending decisions). Increasingly development in Britain is taking place on 'brownfield' sites, to protect the countryside including the green belts established around many of the country's cities and metropolitan areas, and on flood plains. Difficult ground conditions are likely to give rise to greater foundation-related problems in the future. Large programmes of new building in the past have tended to give rise to maintenance problems. For instance many of the prefabs built immediately after World War II have had problems with flaking asbestos; when they were built, problems with that material were unknown. Other buildings built with prefabricated reinforced concrete (PRC) panels in the 1950s and others in the 1960s of large panel systems of construction (LPS) have also proved problematic and many have been demolished or needed to be substantially rebuilt at great expense. Such considerations must be expected to influence maintenance activity on those buildings.

When Ronan Point, a system-built tower block in the London Borough of Newham, suffered partial progressive collapse on 16th May 1968 as a result of a gas explosion (http://news.bbc.co.uk/onthisday/hi/dates/stories/may/16/newsid_2514000/2514277.stm), there was a parallel collapse in confidence in prefabricated constructions. There had long been antipathy toward 'high-rise' living by tenants used to living in terraced housing. Public sector policy was compelled to change direction. Whereas the UK government had hitherto provided subsidy that increased with increasing numbers of storeys, emphasis now switched from comprehensive redevelopment to housing rehabilitation. Government policy switched to the creation of General Improvement Areas (GIAs), Housing Renewal Areas (HRAs), Neighbourhood Renewal Assessment (NRA) and Regeneration.

Despite changes in government and the ups and downs of maintenance and improvement activity in Britain, the political climate is broadly supportive of investment in buildings and fairly constant. That cannot be said in all parts of the world. For instance in the war-torn Middle East or in poverty-stricken Africa, what is the effect of those political climates on building maintenance? In such circumstances it is difficult enough just to keep structures reasonably intact and standing, with water and sanitation services in place. In some other places there are memories of political and economic instability manifested in currency crises and loss of confidence that also may be expected to inhibit investment in maintenance of buildings.

Building maintenance takes place within a legal environment. Hammurabi introduced a legal code in Babylon in c.1760 BC (www.britannica.com), and various codes have been developed around the world since. The system of laws and regulations of the English-speaking world follows largely that of England and Wales, albeit with significant national differences; in continental Europe the base is the Napoleonic Code; in South Africa, Roman–Dutch law prevails.

In the UK there has been quite a proliferation of legislation. A cynic might suggest this has been due to the prevalence of lawyers as members of parliament. Critics complain of tinkering, introducing new or amended legislation before there has been opportunity to really get used to the previous arrangements. Significant areas of regulation relating to building maintenance include:

- employment;
- taxation;
- planning;
- building operations, especially regarding matters of health and safety;
- pollution;
- building performance.

The first two of these, employment and taxation, are strongly linked, and largely beyond the scope of this book. For instance, in the UK, employers are liable to pay to HM Treasury the income tax and National Insurance payments due from their employees under the PAYE (Pay As You Earn) obligations. There are issues of Statutory Sick Pay (SSP) and pensions. There is also Value Added Tax (VAT) to be paid and claimed back. These are areas on which it is sensible to take advice from specialists: accountants, personnel managers (now generally branded as human resource management or HR), solicitors, tax advisors. These issues will, however, impact upon maintenance work. For instance they will determine how much a piece of work will cost, how many staff can be afforded and how far a budget will stretch.

Planning and building performance are also linked to some extent. The UK has had control over land use and development since the Town and Country Planning Act of 1947. This was introduced in order to prevent urban sprawl, the extension of towns and cities out into the countryside. How does this affect building maintenance? Planning legislation controls physical alterations to buildings and changes in use. It may also inhibit, and possibly prevent, the development of a business by opting to prefer and support some other vision of how the area in which the building is situated should develop or be allowed to develop. Planning has developed a strong emphasis on control, stressing what is permitted and, by deduction, what is not permitted. Building owners and architects have for a long time resented such imposition, however well meant and however much it represents the wishes of the local population. Local councillors are dependent on the votes of their electorate to be empowered and enabled to make the changes they deem necessary for the betterment and improvement of their area. 'Nimbyism' reigns (not in my back yard). There have been various moves over recent years to reduce the role of planners in relation to small works, but good intentions still seem to disappear amidst more rather than less red tape. For instance, changes intended to remove more small works from the need to make a planning application are likely to bring forth more control of domestic loft conversions that feature dormer windows (Town and Country Planning Order, 2008).

In the UK, and in countries with similar legislative approaches, government has chosen to take an interest in how building maintenance work is carried out and in the performance of the building afterwards. In relation to maintenance operations the principal interest is in the health, safety and welfare of the maintenance operatives carrying out the work and of those around them using the premises or passing by or through. It is not intended to give an exhaustive history of legislative development here, though the following outline may help illustrate the developing context.

There have long been Factories Acts (from when Workplace and Factory were virtually synonymous, reflecting the previous industrial age), and Construction Regulations. The first Factory Act was established in 1819, largely to prevent abuse of children – it banned children under 9 years of age from working and limited those aged 9–16 to working a maximum of 72 hours a week. The Factories Act of 1961 and the Offices, Shops and Railway Premises Act of 1963 made requirements for the protection and safety of workers and visitors, and placed obligations on building owners and occupiers for the maintenance of the buildings and equipment therein. The Health and Safety at Work Act of 1974 represented a big step forward; it placed obligations on employees as well as employers. This Act set up the Health and Safety Commission (HSC) to regulate health and safety in Great Britain and the Health and Safety Executive (HSE) to carry out research, to give advice, inspect and enforce regulations. HSE employed about 4000 people in April 2006; it maintains an excellent website (www.hse.gov.uk).

The principal regulatory instrument in relation to building operations is the Construction (Design and Management) Regulations. These were introduced in 1994 and substantially revised in 2007 after extensive consultations. They apply to all construction work of over 30 days duration or more than 500 person-days and require the appointment of a 'principal contractor' and a 'CDM coordinator' (see HSE website for details). Construction work includes construction, alteration, conversion, fitting out, commissioning, renovation, repair, upkeep, redecoration or other maintenance (including cleaning). More will be said about the organisation of maintenance work in Chapter 9.

Matters of health and safety in buildings are also subject to public health acts. The first of these was in 1875 and there were major modifications in 1936 and 1961; this latter made provision for the preparation of the first national Building Regulations, which were published in 1965. These swept away a myriad of local by-laws and enabled greater consistency of standards, at least across England and Wales; Scotland had its own Building Standards. There were frequent amendments made, particularly significant being those that followed the oil crisis of 1973 to increase standards of thermal insulation. Some of the regulations were set out in a prescriptive form; others set a performance requirement, exemplified by clauses describing constructional arrangements which, if complied with, were 'deemed to satisfy' the regulation. The Building Act of 1984 heralded a wholesale revamping of the Building Regulations.

The Building Regulations 1985 were issued in a new format with each regulation set out in the form of a brief requirement with accompanying Approved Documents (ADs). The ADs set out ways in which the 'requirements' could be met. Each AD has been revised since first publication. At the time of writing, the oldest currently are 1992 editions of Part D Toxic substances and Part G Hygiene; the most recent (all 2006) are Parts B Fire safety, F Ventilation, L Conservation of Fuel and Power, and P Electrical safety. Each AD is prefaced with a statement to this effect: Approved Documents are intended to provide guidance for some of the more common building situations. However there may well be alternative ways of achieving compliance with the requirements. Thus there is no obligation to adopt any particular solution contained in an Approved Document if you prefer to meet the relevant requirement in some other way.

Prior to 1985 Building Regulations approval needed to be sought before the commencement of controlled works: it was necessary to apply and to comply. That process, known as a full plans submission, is still possible and most appropriate for large works. It is not, however, mandatory to go to the local authority for approval; it is possible to go to an approved inspector licensed on behalf of the UK government by the Construction Industry Council. It is also possible to use a Building Notice process, more suitable for small or repeat works, where the client or his/her agent gives notice of the commencement of works. It is then for the building inspector to arrange to visit to view the works at appropriate times to ascertain their compliance with the Building Regulations.

The Building Regulation system, however, only assures compliance at the time of design and of construction. There is no obligation to keep improving a building so that it meets the regulations that would apply if the building were being built at that time. Nor, until recently, has it been necessary that any improvement work being planned or carried out should conform to the new standard, only that the building, or part, should be made 'no worse'. However, while it remains generally the criterion, that is not the case where windows are being replaced. In this case the new windows need to meet the standard required of new windows; that is to say that single-glazed windows being replaced will need to have double- or triple-glazed units substituted. There is sense in this though it does challenge the general 'no worse' expectation. There has been consultation on the possibility of requiring all work to comply with current standards. The concern is that improvements may not be made at all if the cost of work becomes too great. A further possibility would be to require all buildings to be continually improved to meet the standards of the day, rather like the MoT test applied annually to motor vehicles over 3 years old. There is currently no requirement to maintain a particular level of building performance, other than in the matter of energy performance.

The Energy Performance of Buildings Directive (EPBD) (http://ec.europa.eu/energy/demand/legislation/buildings_en.htm) was promulgated by the European Union in 2002 (COM 2002/91/EC) (European Union, 2002) in response to the 1997 United Nations Kyoto Protocol. Energy Performance Certificates (EPCs) are required when buildings are constructed, sold or rented out; and Display Energy Certificates (DECs) are required for 'large, public buildings occupied by public authorities or institutions providing a public service to a large number of persons, total useful area greater than 1000m^2' (www.communities.gov.uk/planningandbuilding).

EPCs are a mandatory part of the Home Information Pack (HIP) now required to be provided by sellers of houses and flats in UK. There was a period of confusion around the introduction of HIPs that resulted in the omission of the proposed Home Condition Report. This would have given buyers insight into the maintenance and repair needs of their intended purchase, but this fell due to uncertainties about the legal status of the report in relation to the normal expectation of *caveat emptor* (let the buyer beware). The extent to which buyers and lessees may rely upon EPCs and HIPs is yet to be seen or tested in the courts.

In Australia, a performance-based national set of building regulations (Building Code of Australia) came into force in 1996. The regulations are set out and described at four levels: (www.abcb.gov.au):

- Level 1   Objectives
- Level 2   Functional statements
- Level 3   Performance requirements
- Level 4   Building solutions
- Level 4a   Deemed-to-satisfy provisions
- Level 4b   Alternative solutions

In the UK, in addition to there being no need to maintain a building to any particular standard, there is no requirement for maintenance work (or for that matter any building work) to be carried out by skilled operatives. Britain has become used to the scourge of the so-called 'cowboy builder'. People would be horrified to fly in aircraft built or maintained by operatives without the relevant skills or qualification so why is that tolerated in buildings? In the 1990s this was addressed by a government task force under Tony Merricks, MD of Balfour Beatty (Department of the Environment, Transport and the Regions, 1998) and resulted in the introduction of the Construction Skills Certification Scheme (CSCS) to be administered by the then Construction Industry Training Board (CITB, now ConstructionSkills). This is discussed in more detail in *Building Care* (Wood, 2003, pp.2–3).

The Main Contractors Group was hoping to have a fully qualified workforce by 2007; it is currently hoped to have a fully qualified workforce by 2010. Part of the problem stems from the cyclical nature of construction, with highs and lows of construction work affected by the state of the nation's (and the world's) economy, public sector programmes and private investment. This makes employment prospects and career planning uncertain. It is not surprising, therefore, that construction is well below law and media studies as a career choice. Construction has an unattractive image. The industry has failed to attract women into it. Despite the satisfactions available to a skilled bricklayer, joiner or stonemason, building maintenance is not 'sexy'. The market has responded to this unmet need by the 'supply' of relatively cheap skilled labour from recent additions to the EU such as Poland and the Czech Republic. The EU offers and promotes free mobility and migration of labour. In Australia the response has been to require 'builder registration' in some states (e.g. Victoria) rather like the UK's CSCS card (Georgiou *et al.*, 2000).

Finally, regarding political influences, we cannot afford to ignore internal politics. This is about recognising and understanding who it is that gets his or her way, and how and why. Is it about 'who shouts loudest', or longest – decibel-determined maintenance? As a maintenance manager you will want, and need, to be sure to get your way, especially on important matters. More will be said on maintenance planning in Chapter 3. Those who influence or control your budgets and programmes will need to fit your plans and aspirations into theirs, or to slim down or reject yours. It may be useful to construe your plans into essential work and desirable work. Work to maintain, or restore, health and safety must be done – that is a legal requirement. For other work you need to show how it furthers the objectives of the organisation: improves working conditions and productivity, demonstrates corporate social responsibility, will improve the environment, enhances image, shows a great return on investment, has a short payback period, etc. The ability to 'talk money' will be invaluable, especially to convince those who 'hold the purse strings', or 'bean counters' as they may sometimes be called, rather pejoratively.

Economic

Aspects to be considered here include:

- Macro-economic factors, such as the value of a building, and its enhancement or reduction by improvement or decay and obsolescence; the national and international economic contexts; the use of construction programmes, their promotion and prolongation (stop/go) as an economic regulator.
- Micro-economic considerations, such as costs of labour – wages, National Insurance, sick pay, recruitment, severance costs; subcontracting and self-employment; outsourcing.
- Materials costs and availability; discounts for bulk purchase.
- Hiring or buying of construction plant and equipment.
- Profit and loss; waiting and unproductive time.

For many people, this is the over-riding imperative: the bottom line. Although there is mention occasionally of a 'triple bottom line' (Elkington, 1998) this struggles to overcome the simplicity of the single criterion. An advantage of the bottom line approach is that it can be calculated. A lower figure on expenditure or a higher return on investment are self-evidently better. This can of course be an over-simplification, but it does make for a baseline against which to discuss alternatives. For instance, it may make sense to spend more than the minimum on a maintenance task to provide a (hopefully) long-lasting replacement component than to make a poor quality repair. A life cycle analysis could be undertaken and a calculation made to determine the lowest through life cost. This will be developed further in Chapter 5.

At an operational level, the maintenance manager, or someone on his or her behalf, will need to make assessments of how much fellow professionals and operatives will be paid. This takes place against a political background of, for instance, a national minimum wage, industry norms, supply and demand. From time to time, governments may change tax rates, whether on income or expenditures (e.g. the increase, application or removal of Value Added Tax), pensions or National Insurance.

Such imposts may influence the manager's ability or desire to arrange maintenance work in a particular way. For instance approved alterations can be zero-rated for VAT when carried out on a protected building. Approved alterations are ones which cannot be carried out without Listed Building Consent (and have received that consent) but do not include repairs or incidental alterations made as a result of the required repairs. A protected building is a listed building which is to remain as or become a dwelling, or is intended to be used for a communal residence or charitable purpose after the works are carried out. Perhaps improvement works on student rooms at an Oxford or Cambridge college might meet those criteria; it may be worth having a taxation specialist on board to help with such considerations.

In Britain since World War II there has been a decided move away from direct employment toward more self-employment, labour-only contracting and subcontracting, and outsourcing, where the procurer of the service, in this case the building owner has no responsibility for the employment of staff to carry out the maintenance activity. This has come about because of the costs and uncertainties of being able to keep staff continuously and fruitfully engaged when budgets and therefore workloads fluctuate. It does however mean that some of the benefits of continuity are lost, such as the 'corporate memory' of what has been done in the past, and how and why. That may be thought unimportant if good records are maintained, but often they are not. There may also be losses in quality of work done where skills are no longer available, and delays where there is no labour available at all to do the job at the desired time.

In similar vein, materials need to be available in the right quantities in the right place, at the right time and at the right cost. A good maintenance manager may be able to negotiate good prices for large or repeat orders, but there may be limited scope for the long-term storage of materials. There may be benefit in seeking an over-ordering at the time of new or major works in order to provide a small stock of matching material to use for repairs. This will only be of value, however, if the stored material weathers at the same rate as that incorporated into the works originally. In repair and maintenance work there is often less scope than hoped for such matching, and it is often better to recognise that matching in is not possible at any cost and to accept that the new work will not blend in. This is ultimately an aesthetic decision.

The maintenance manager may find himself or herself needing to consider the hiring or purchasing of construction plant and equipment. There may be work that needs extensive scaffolding or a cherry-picker to reach a point which is high up and hard to reach. The main considerations here would be the size and scope of the work, its duration and the level and type of skill needed for erection and dismantling or for operating equipment. In general, frequent and repetitive activities are likely to suggest purchase of such equipment and training of staff with the requisite skills; infrequent, one-off or specialised activities suggest the hiring in of appropriate plant with associated skilled operatives. A cost comparison could be made, though the answer arrived at would be strongly influenced by the variables put in to the equation.

Labour is the greatest cost of almost all construction activity in UK and especially so in the case of RMI activity. The management of the staff resource is likely to be the biggest contributor to profit or loss on a project. The minimisation of waiting and unproductive time is a key component. Unfortunately RMI work includes a lot of taking down of existing work and the revealing of conditions worse than expected, with consequential changes to the work to be done. Changing plans always takes time and costs money. There is often further investigation needed, additional work, or work to be done out of sequence, people awaiting instructions, others held up waiting for completion of work by preceding trades or for drying out. It is prudent to build in substantial contingencies for extra time and costs.

Pricing books and schedules of rates are available for the estimation of costs and times for minor works, repair and improvement works from such as *Spon* and *Wessex* and available in software packages as well as the traditional hard copy. Such 'formulaic' approaches may have their limitations – maintenance actions on old buildings can be so individual and differ from the standard activity described. The 'swings and roundabouts' argument pertains; over- and under-estimates tend to even each other out. What may be more insidious is a tendency to build budgets on last year's expenditure +/– a percentage, up for inflation, down for efficiency gains. This fails to take account of actual needs of the building, as discerned from condition surveys, and of organisational needs, as suggested by a corporate strategic overview.

Social

In some parts of the world there may be social or cultural constraints on who may be employed; family ties may be thought appropriate and beneficial or may be seen as inappropriate and unadvised – nepotism, cronyism, old school tie. In South Africa there have been moves to more labour-intensive construction. This makes sense when there is a large pool of labour that may be unemployed, idle and indolent, and high technology, materials and equipment are expensive.

There may be aspects of effective working by associating operatives in mutually supportive teams, although there may be economic downsides if not everybody is fully occupied with productive work all the time. In UK, although work may generally be seen as a means to an end ('I work to live, rather than live to work'), this attitude can tend to devalue work. Certainly when people become unemployed they often find out how much they then feel de-valued, their self-esteem radically challenged, no longer needed, 'on the scrap-heap'. Many people describe themselves by what they do – that is who they are. Often a person's closest friends are his or her work colleagues. Effective work-teams can be very much self-managing, giving mutual support and the satisfaction of doing a good job and enjoying the doing of it.

Sometimes work can be become all-consuming, especially when one is confronted regularly by new challenges repair and maintenance work often throws up. The economic imperative to keep everyone fully engaged and fruitfully employed puts pressure on managers to make decisions quickly, and for those decisions to be the right ones; this can be stressful. Managers may take their work home with them (whether physically or metaphorically) in the evenings and at weekends. It can be hard to switch off. Managers, including maintenance managers, need to keep this in mind, to keep themselves and their colleagues under review and in check. Britain 'enjoys' a high divorce rate. Don't let it happen to you; set achievable goals; go home at a reasonable time; maintain a healthy work–life balance. Consider whether your organisation would benefit from showing its employees their value through seeking something like 'Investors in People' status or similar. It is important to have and show respect for people. In some parts of Britain there are protocols and partnerships entered into for the employment of a percentage of local labour. This helps lift some people out of unemployment and poverty and helps instil in them pride in themselves and in their neighbourhood and communities, which they are simultaneously helping to improve.

The relative labour-intensity of building maintenance and repair work has great potential for the regeneration not only of run-down areas but of run-down people too.

Technological

All buildings decay and deteriorate over time. Some materials deteriorate more quickly than others, and more quickly in some situations than others. Such deterioration is not, in itself, a problem – but it is a problem when that deterioration reaches a point of unacceptability. It may be that it fails to meet a performance expectation, or that it is visually unacceptable, or perhaps becomes dangerous. To an extent an expected rate of deterioration can be forecast, from data related to similar materials in similar locations at times in the past. That has its limitations, and more will be said on this, and about deterioration in more detail, in Chapter 7. Defects (situations in which parts of the building have failed to be acceptable) are illustrated and discussed more thoroughly in Chapters 10–15. Many defects arise from situations that could have been identified, and dealt with, at the design stage; others stem from poor standards of workmanship, resulting in failure much sooner than might have been forecast or expected.

I will give here just two brief examples of problems related to material deterioration:

▪ Timber – a material used down the centuries with relatively well known and understood properties. It is also a material of the present and future; potentially a renewable resource, with good sustainability credentials. However, in situations of too much or too little moisture it is prone to rot and decay and to cracking and distortion. It is also readily attacked by insects – weevils, beetles, termites – often with fearsome sounding or Latin names, such as death-watch beetle or *Xestobium rufovillosum*. Deterioration may be gradual; it may be unseen; it may possibly be catastrophic.

■ Metals – their use increased greatly in the Industrial Revolution, from nails to long spans in steel. But problems arise where different materials are used in proximity, especially in damp conditions. Electrical currents are set up and electrolytic (or galvanic) actions take place, causing corrosion. Some metals (such as zinc and aluminium) will thin down and fail; this can de dangerous if, for example, they are fixings of structural elements, or if intended as a protective covering (as in galvanised bolts or a water tank). And some metals in some timbers can give rise to corrosion problems too; for instance oak, western red cedar or Douglas fir may corrode galvanised nails.

So even well known, tried and tested materials fail. How should maintenance managers respond? Should we encourage more use of new materials and methods? Generally the response to problems is to retreat to the more conservative, to be more certain to get it right. Certainly in relation to the maintenance and repair of historic, listed and heritage buildings, the expectation should be to try to perpetuate the original design and construction. This will usually involve trying to keep as much of the original material in place. How much cross-sectional loss could a structural member sustain and remain safe? Improved technological understanding may help here. In the past, materials were sized on the basis of what had been observed to work or to not work. For instance many of the magnificent Gothic cathedrals of northern France had parts rebuilt several times as their stonemasons strove to build higher and higher:

> Beauvais Cathedral was never completed westward of the choir and transepts... The roof fell (1284), and the choir was reconstructed and strengthened by additional piers (1337-47)... There was an open-work spire, 500 feet {150m}high, over the crossing, which collapsed in 1573... The building is of extreme height, 157 ft 6 in {48m}to the vault... This soaring pile is perhaps the most daring achievement in Gothic architecture... The structure is held together internally only by a network of iron tie-rods, which suggests that these ambitious builders had attempted more than they could properly achieve. (Cordingley, 1961, p.542)

We now have structural engineers and software to analyse the structural dynamics of the building and how the forces are distributed through the structure. We may find that the member in question is largely redundant from a structural point of view and that repair work could be seen as principally aesthetic. Or, of course, the member may be vital and the structure in some peril. Perhaps a strengthening or rearrangement elsewhere in the structure may provide sufficient relief. Perhaps new, stronger, lighter materials may help. Personally I am happy for a twentieth or twenty-first century material or method to be incorporated into an older building, but that is a philosophical matter and I recognise that mine may not be a majority view. I will say more on this in subsequent chapters.

Research and development (R & D) in the construction industry is not generally high on the agenda. In a response to a parliamentary question on 21st March 2007, the then Secretary of State for the then Department of Trade and Industry, Margaret Hodge, is recorded in Hansard (2007) as stating the value of 'construction contracting R and D' as 0.06% of the gross value added. She added that 'it is difficult to capture innovation in project-based activities such as construction'. If construction output is valued at £81 billion (Office of National Statistics, 2007) that would suggest a figure of about £48 million a year. The Building Research Establishment was reported (BRE Trust, 2007) to have spent just under £40 million in 2006. While much may be spent, rightly, on researching modern methods of construction, comparatively little is spent on research into building maintenance and repair. Bearing in mind that existing buildings offer the greatest scope for reducing energy use and therefore carbon emissions and global warming, it would be good to see more funds allocated to research in relation to existing buildings and their upgrading in terms of performance. I will say more about this in later chapters.

2 Design temptations

The previous chapter has set the scene in relation to building maintenance. The purpose of maintenance has been considered and definitions explored. The scope and scale of maintenance activity have been examined. From the foregoing it can be seen that maintenance will have a big influence on the sustainability of a building as well as the effectiveness of operation of buildings and what goes on in them. The strongest determinants of future building maintenance are the building's design and construction, its past and current maintenance, and how maintenance is perceived to relate to operational efficiency. This chapter focuses on the critical effect of design.

Buildings convey something of the spirit of the age and of place. Figures 2.1–2.12 show something of the development of London. In Figure 2.1 the eleventh century solid stone Tower of London is contrasted with the twenty-first century steel and glass City Hall across the River Thames. Figure 2.2 illustrates the development of the financial heart of the City of London. It shows the classical nineteenth century Bank of England (Sir John Soane, architect) and Royal Exchange buildings (Sir William Tite, architect), with twentieth century buildings behind at 152 Broad Street (the site of the former London Stock Exchange building) and Tower 42. The Royal Exchange formerly housed Lloyd's coffee house that was the centre of business in the eighteenth century. Tower 42 was built as the headquarters of the then National Westminster (NatWest) Bank. Constructed in the property boom of the 1970s to a design by Colonel Richard Seifert, and refurbished and reclad in the 1990s by GMW architects, it was for a time the tallest building in the UK. Figure 2.3 shows part of the recent Paternoster Square redevelopment in the shadow of St. Paul's Cathedral with brick buildings replacing the no longer fashionable concrete structures and windswept piazza of the 1960s. Figure 2.4 is of Canary Wharf, an area redeveloped in the 1980s from disused and derelict docklands. The government of Mrs (subsequently Baroness) Margaret Thatcher introduced Enterprise Zones to encourage and facilitate economic revival in rundown areas such as the London Docklands, relieving local authorities of their development control powers. One Canada Square (Cesar Pelli, architect) was constructed as the tallest UK tower (superseding Tower 42). Developers Olympia and York went bankrupt in the process.

Figures 2.5–2.8 show further examples of changes in use of city space. Figure 2.5 shows St. Katherine's Dock, adjacent to the Tower of London, where some former warehouses were converted in the 1970s to offices and a convention centre, and others replaced by a hotel and leisure facilities related to a new marina (Renton, Howard, Wood, architects). Figure 2.6 shows the former Billingsgate Fish Market, which relocated to southeast London. Figures 2.7 and 2.8 show developments on former railway lands, the site of the former Broad Street terminus (adjoining Liverpool Street) and of Somers Town goods depot. The former is now the Broadgate

Figure 2.1 The Tower of London and (across the River Thames)London City Hall.

development, built using 'fast track' techniques in the 1980s by Rosehaugh/Stanhope Properties and British Rail, providing 4.8 million square feet of commercial floorspace; the latter is the British Library (designed by architect Sir Colin St. John Wilson in 1962, and completed in 1997). The British Library is overshadowed by the spires and turrets of the nineteenth century former Midland Grand Hotel (Sir George Gilbert Scott, architect) of St. Pancras Station, which has been transformed in the twenty-first century to become the London terminus of Eurostar trains to

Figure 2.2 The financial heart of the City of London.

Figure 2.3   Paternoster Square, adjoining St. Paul's Cathedral, London.

Paris and Brussels. The economy keeps moving on, making new demands of our buildings and facilities.

Most commercial buildings will have been designed by architects, or more accurately, within an architect's office. Although the 'typical' architect's practice is that of a sole practitioner, most architectural practices will operate with assistance from architectural technologists. The architect will have the meetings with the client, and prepare the first 'back of envelope' sketches that translate into some kind of design – that is to say, what the architect understands the client to

Figure 2.4   The towers of Canary Wharf dwarf London's 'East End'.

Figure 2.5 St. Katherine's Dock, London – now a marina.

require. It is likely that an architectural technologist will then have the responsibility of working up the details.

Note: Although the title 'architect' is protected by law in UK (Architect's Registration Acts of 1931 and 1938 and the Architects Act 1997), the practice of 'architecture' and the use of the word 'architectural' are not protected.

By contrast, most residential buildings other than single, one-off, designs are more likely to have no involvement of architects. Typically they will be designed by architectural technologists

Figure 2.6 The former Billingsgate Fish Market, London.

Figure 2.7　The Broadgate development, adjoining Liverpool Street Station, London.

or chartered surveyors or surveying technicians. Most housing estates will be designed from a 'pattern book' of repeated designs or 'house types'. These are typically given names that conjure up a high-class imagery, such as the Blenheim or the Burleigh, rather than the Bradford or the Brixton. In the case of social housing (which used to be known as council or local authority or housing association housing), these are more likely to be designated House Type B1 or B2.

Architects may be involved in designing building conversions and adaptations, although again these are more likely to be entrusted to building surveyors. An understanding of structures

Figure 2.8　The spires of St Pancras Station overshadow the British Library.

Figure 2.9 Tower 42, formerly the NatWest Tower.

and how loads may be transferred if the building is in poor condition or if walls are to be removed is important. The education and training of surveyors normally includes material on the maintenance and modification of buildings in addition to their construction. By comparison, the education and training of architects, although it is longer in duration, is generally lighter in these aspects. Architecture students are more commonly attracted by the possibility, perhaps imperative, of changing the world through their designs. This can induce or encourage a tendency to the one-off, eye-catching, iconic design.

Figures 2.9–2.12 show iconic buildings of their day. Figure 2.9 shows the 1970s designed and 1990s reclad Tower 42; Figure 2.10 is of the 1980s Lloyd's of London (Lord Richard Rogers, architect) with its external services, including glazed lifts. Figure 2.11 is of the twenty-first century 30 St. Mary Axe, commonly known as 'the Gherkin' (Sir Norman Foster, architect); while Figure 2.12 shows the 1970s suburban 'brick-and-tile' Hillingdon Civic Centre, Uxbridge (Robert Matthew Johnson-Marshall, architects).

I am focusing somewhat on these professional differences, even caricatures, not to pillory fellow members of the team, but to encourage better understanding, acceptance and even delight in the rich diversity of inputs available. But when things go wrong it is understandable that there may be a tendency to seek someone to blame rather than to focus on putting matters right. The client will certainly not be to blame – after all he or she is paying! However, they are most likely to be the least experienced and knowledgeable in either knowing what they want or need or being able to express it effectively. Of course it would be better to avoid things going wrong in the first place – or would it?

Figure 2.10   Lloyd's of London.

Design standards

How are appropriate standards determined, and by whom? The word 'standard' may be unhelpful; it is ambiguous in meaning. It could be used in relation to a repetitive way of doing things: 'that is the standard approach'. Alternatively, it can be used in the sense of a minimum level of performance to be achieved. It is in this latter sense that it is commonly encountered in, for example, Building Regulations or as defined by British Standards. There are also other national standards, European Norms and those of the International Standards Organisation. There is a sense in which these standards represent the lowest common denominator, the lowest acceptable performance, rather than the best achievable. There is a process in Europe of harmonisation of standards, in the quest for elimination of non-tariff barriers to free trade. National standards have developed in the context of national cultures and practices ('the way things are done round here') that also reflect local climatic conditions. The harmonisation process questions and challenges the validity of these rationales and seeks out common ground. National representatives could perhaps be forgiven for trying to persuade others of the virtue of their way. It is against EU rules to demand one's national standard be applied to the exclusion of work that complies with the standard of another EU country. For instance Spanish standard drains should be acceptable in Ireland. Suffice it to say here that the vast majority of the buildings with which we will be concerned will have been designed to the standards of the country within which they are located using materials that conform to that same nation's standards. Similarly the materials, products

Figure 2.11　30 St. Mary Axe, London ('the Gherkin').

and practices most available and likely to be used in repair and maintenance work are also likely to be those of the area in which the building is located.

There can be dangers in applying inappropriate standards. For instance, to apply to a building on the Spanish Costa del Sol design standards appropriate to a building north of the Arctic Circle in Finland would be to ignore important climatic differences that may severely affect the building's performance. Similarly to replace a window in a bungalow in Greece with one designed and tested for a 40-storey building in Singapore (or vice versa) may be to grossly over- (or under-) provide for the situation. Although there is in Britain a notional design life of 60 years, the life that may be reasonably expected of individual components is not well understood. Information on actual lifespans is hard to find. The 60-year life stems from the repayment period of loans from the UK government's post-war Public Works Loan Board. It was reasonable to expect that buildings on which public money was spent should last at least the length of the loan period, if not substantially longer. If only ... Some of the tower-blocks and other mass housing solutions to the housing shortages of the 1960s and 1970s did not last even half that period.

So what is a reasonable life expectancy for a building or component? What then is a reasonable standard? I suggest that more time and effort should be put to considering, and answering, these questions at the design stage. To design for a notional 60-year life would be to over-design if it could be known that the building is going to be demolished in 30 years. Bearing in mind the frequent talk of an ever-increasing rate of change, it may be prudent to design for a 20- (or

Figure 2.12 The 'brick-and-tile' aesthetic of Hillingdon Civic Centre, Uxbridge.

30- or 10-) year horizon. Who can forecast what will be required of our buildings? This and related issues will be explored further in Chapter 5.

As an architect myself, I venture to suggest that many in that profession have the abilities and propensity to look ahead and to consider a range of possible futures. Some others involved with buildings may see such people as 'air-heads', which runs the risk of missing out on creative thought at a stage when it could make a significant impact on decision making. This is not to say that 'hair-brained' schemes should be pursued, only that alternative ideas should be considered and evaluated at the design stage. It may even be that the best solution to the client's 'problem' is not a building but perhaps a change in processes or priorities. I would go so far as to say that it is the responsibility of the design team to put before the client the broadest range of conceptual possibilities before homing in on the preferred approach.

Design team

Who is in the design team; and why is a team needed? In short, no one person is likely to have all the skills needed to enable the best decision to be made. Buildings are more complex entities than a mere pile of bricks and mortar. This is especially so after they have been in use for some time, with deterioration set in to varying degrees and with repair, maintenance and improvement actions already taken at various times. They are likely to have parts which have moved over time, whether due to subsidence or other factors, and other parts, particularly in relation to building services, which have become worn out and obsolescent. A mature and holistic view is required.

The following skills could be useful to have when considering and deciding upon significant repair, maintenance and improvement plans:

- someone who was involved with the original design/construction;
- quantity surveyor;
- building/maintenance surveyor;
- facilities manager for the building(s) in question;
- structural engineer;
- building services engineer(s);
- client.

It is likely that there is nobody available who was involved with the building originally. It used to be that it was an offence under the Royal Institute of British Architects (RIBA) code of professional conduct to supplant a fellow architect. It was deemed appropriate that the original architects should be approached if it was intended to design work to alter or extend a building they had designed. Notwithstanding the ethics of the situation, it makes sense to make use of the knowledge residing in those who were involved at the time. If nobody is available it may be that at least drawings and other documents related to the building's construction can be located and consulted.

A quantity surveyor (QS) is essential for the estimation of costs and their role in the evaluation of design alternatives. I would not expect the QS to provide aesthetic judgement but their contribution regarding economics will be indispensable. Their ability to 'talk money' with the client will be invaluable. QSs also tend to be able to offer good advice on selection of forms of contract and procurement generally.

It may be that you (the reader) will find yourself in the role of building surveyor or maintenance surveyor. You may be on the staff of the existing or proposed building owner or occupier, or you may be engaged as a consultant, either just for this project or on a long-term basis. You will be expected to have, or to obtain, a familiarity with the building, its features and its state of repair. A chartered surveyor is one who is a member of the Royal Institution of Chartered Surveyors (RICS).

You may be the facilities manager (FM) of the building(s) in question. As such you will be responsible for the operation of the building and what goes on in it. Because the backgrounds and specific responsibilities of FMs are so disparate, you may know little about the construction or maintenance of buildings; or you may know so much as to assume the role of the building surveyor or maintenance surveyor. Either way, you are going to have a continuing responsibility for the care of the building into the future, so it is important for you to be involved in the decision making from the outset.

A building surveyor may know a great deal about structural behaviour of buildings, and he or she may feel qualified to make assessments of the stability and adequacy of a building structure. It is well to be sure that a person claiming to have that skill has a professional indemnity policy in place that covers that; if not, a structural engineer should be engaged. They are less likely to be needed full time and it would be usual to call upon his or her services as and when required.

All but the most basic of buildings will warrant the assistance of a building services engineer. Depending upon the scale of the operation more than one may be required. Some engineers specialise in one service or another; some firms offer comprehensive coverage: heating, ventilation, air conditioning, electrical, communications, lift installations, fire and security systems, the list is long and lengthening with growing use of increasingly complex technologies.

And last but not least, the client must be fully engaged. The day when the professionals knew best (if it existed) has passed. The client is the one who is paying, who is using or about to use the building or is going to sell it on or rent it out. He or she must make the decisions, albeit based on the available advice.

Each of the professionals involved should have a contract in place with the client that sets out clearly the scope of their work, terms and conditions of service and arrangements for payment and for varying the contract.

Who leads the team may be as important as who is in it. The typical client only builds once – whether that relates to the bad experience that many may have of dealing with construction projects and personnel may be a moot point. Certainly most people do not have need of more than one building and that may be a pre-used one. However, many people do move house several times, and most will commission repair, maintenance or improvement work. In that sense many people will be amateur clients at some time. Those involved with the determination or recommendation or execution of such works may find themselves working with many first-time clients. Working with professional or repeat clients may be comparatively rare; in these cases it may be appropriate that the client should control the project, coordinate the various professional inputs and chair meetings. In other situations it may be best for one of the professionals to lead the project and to be responsible for all communications between the team and the client. Traditionally for newbuild projects this would be the architect; for work on existing buildings it may be a chartered surveyor, whether a QS or building surveyor. Where a FM or maintenance manager is employed by the client it may be appropriate for him or her to lead and/or to take the role of client's representative, making the decisions or acting as a conduit into other parts of the corporate client's organisation in the pursuit of a decision.

People are, of course, more than a role. It is not appropriate to suggest, for instance, that all architects are the same; they are not, nor are all surveyors, etc. Many tomes have been written on team roles and effective teamwork – too many to summarise here. Airport terminals and railway station bookshops are full of advice on how best to lead or manage other people to get the best out of them by adopting the style of some military hero or business guru. I listed in *Building Care* (Wood, pp. 49–50) a 'galaxy of gurus'; I will mention here just a few that relate to teamwork:

- Armstrong, M. (1994) *How to be an Even Better Manager* (4th edn).
- Belbin, R.M. (1981) *Management Teams: Why They Succeed or Fail.*
- Blanchard, K. *et al.* (1994) *The One Minute Manager Builds High Performing Teams.*
- Covey, S. (1989) *The 7 Habits of Highly Effective People.*
- Katzenbach, J. & Smith, D. (1993) *The Wisdom of Teams.*
- Norton, B. & Smith, C. (1998) *Understanding Management Gurus in a Week.*
- Roberts, W. (1987) *Leadership Secrets of Attila the Hun.*
- Skinner, Q. (2000) *Machiavelli: A Very Short Introduction.*

From studies at Henley Management College, Belbin identified eight roles necessary for a successful team, which he called:

- company worker;
- chairman;
- shaper;
- plant;
- resource investigator;

- monitor–evaluator;
- teamworker;
- completer–finisher.

Belbin did not suggest that therefore the ideal size of team was eight; some people could fulfil multiple roles; and people need not fulfil consistently the same roles in different teams. For instance, although the architect may have the best design and conceptual skills, he or she may not be the best of those in the team at evaluating proposals and gaining consensus support. It is the team *roles* that need somehow to be filled. Belbin provided a 'self-perception inventory' for people to be able to identify their own best or preferred team role and those to which they are also well suited. Each has 'positive qualities' and 'allowable weaknesses'. The roles and typical features as described by Belbin are summarised as follows:

Company workers are diligent, disciplined and dutiful, but perhaps inflexible and closed to unproven ideas. The *chairman* is calm, confident and controlled, not particularly creative. The *shaper* is said to have drive and dynamism, but may get irritated and impatient. The *plant* is imaginative, individual, perhaps intellectual, thus with his or her head in the clouds and uninterested in detail or protocol. *Resource investigators* are typically energetic, extroverted and enthusiastic, but that may quickly pass once the initial excitement has waned. The *monitor–evaluator* may be thought to be unemotional and hard-headed in applying judgement; possibly de-motivating. *Teamworkers* are social beings, mild, sensitive; they promote team spirit but may be indecisive in a crisis. The *completer–finisher* is painstaking, perhaps a perfectionist; he or she may worry about details and find it hard to let go.

Designs and plans need to be prepared with insight and imagination in their early stages and submitted in their development to critical appraisal and review before being finally accepted with whatever further refinement may be suggested along the way. As has been attributed to Clausewitz and other military leaders: 'No plan survives intact the first engagement with the enemy'. Now it may be a little harsh to describe one's fellow team members as 'the enemy' but it is important to be able to give and receive criticism in the quest for improvement. (I don't take criticism well myself; it's a shortcoming of mine.) More will be said about teamwork and its application to the development and execution of maintenance in the chapters that follow.

I suggest it is important for FMs and maintenance managers to be involved at the earliest stage of consideration of proposals for a new building or improvement of an existing building. This is when there is greatest scope for influencing the amount and pattern of expenditure likely to be needed on the building over time. There is nearly always, but not always, pressure to keep down the initial cost of the building, and to let the future look after itself. A counter-view exists that it is worth spending more on the initial building in order to save money in the long run. Neither view is universally true; both are supportable, depending upon the circumstances. Sometimes a building may be designed that is so one-off or iconic, perhaps using relatively innovative, and untested, materials or constructional form that both its initial cost and cost in use are difficult to predict. Much will be influenced by decisions taken about the expected life expectancy and performance of the building and its components. Who has the knowledge and skill to inform or take such decisions?

The facilities or maintenance manager is unlikely to be involved much, if at all, in questions about the building's 'buildability'. Often building contractors are also little involved at the design stage and have to try to work out how to construct the building when presented with the drawings of how the finished product is expected to look. It would be good for both to be involved. Not only do contractors need to be able to put together the components so that they are wind- and weather-tight but also within the designed and manufactured tolerances (the amount by which they are acceptable over or under the designed size, or out of shape). They also need to

Both focus on the vital importance of an analytical approach. In a similar vein, Keniche Ohmae, from McKinsey's management consultancy and influential in the growth of Japanese industry, writing on the development of a strategy described the strategist's method as being 'simply to challenge the prevailing assumptions with a single question – why? – and to put the same question relentlessly to those responsible for the current way of doing things until they are sick of it' (Ohmae, 1982). It is this fundamental questioning that is the key to the development of both good strategy and sound plans based upon it.

The development of strategy is a high-level activity. It will be done infrequently, although it may be reviewed and revised from time to time, perhaps periodically and especially in the light of changing circumstances. If we take the OED definition of strategy as 'a plan designed to achieve a long-term aim' we can discern some components of what needs to be attended to in its development:

▪ determination of an aim;
▪ a focus on the long term;
▪ a design process;
▪ assessment of means to secure achievement;
▪ arrival at a plan.

In a circumstance where there is already a strategy in place, the process will be slightly amended:

▪ reconsideration of aim in the light of circumstances;
▪ a focus on the longer term;
▪ a review and redesign process;
▪ evaluation of how well the existing strategy has been delivered; processes to change – strengths to build on, weaknesses to eliminate or overcome;
▪ an amendment or addition to or extension or replacement of the existing plan.

Taking the situation where strategy is to be developed *ab initio* (from the beginning), it is important to seek, welcome and weigh inputs from a range of contributors. The organisation's overall strategy will be crucial to it success, particularly in a competitive environment; it needs to be well thought through and agreed so that all are working with the same ends in mind. A wide range of stakeholders, including users and those responsible for the management and maintenance of buildings, should be involved in the production of a strategy for development and use of the estate. The need for buildings stems from the organisation's overall strategy, and any strategy for the maintenance of buildings will be subservient to that. Building maintenance is done to serve the purpose of the organisation, whatever that may be. Figure 3.1 shows diagrammatically how the views of a range of stakeholders may feed into the development of an estate strategy.

Even in a small company the chief executive (or whatever title he or she may have) will benefit from being able to 'knock ideas around' with others. There may be fundamentally different thoughts of what the organisation exists for. Even if an objective has been agreed, for example maximising profit, there may be several ways of seeing it, such as:

▪ maximum net amount of profit;
▪ maximum percentage return on investment;
▪ minimising costs of production;
▪ maximising the share or size of market;
▪ securing a niche market of high quality;
▪ establishing a firm foundation for subsequent growth.

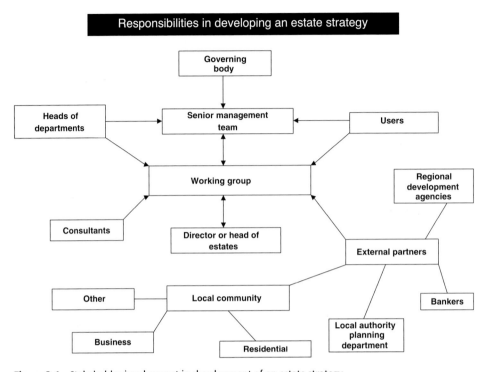

Figure 3.1 Stakeholder involvement in development of an estate strategy.

Of course there will also be several alternative routes to reaching each of those objectives. And there are certainly more objectives than just maximising profits. If, for instance, that were the sole purpose of an organisation, it may be that it would change from one manufacturing widgets to being a specialist provider of healthcare services if there were no profit in the former and it was assessed to be excellent in the latter. Staff could be sacked or retrained and buildings could be sold or let or reconfigured – everything is possible, though not necessarily feasible or desirable. Many companies went through processes of diversification and return to core business, merger and demerger, delayering and downsizing through the 1970s, 1980s and 1990s. These changing priorities had significant impacts on company's needs for buildings.

Consolidation tends to bring need for fewer but larger centres of manufacture and administration; decentralisation and diversification lead to a wider spread of smaller premises. A perceived need for an organisation to be agile and able to adapt quickly to changing markets, methods and priorities might suggest a more flexible approach to property. It is also likely to result in less enthusiasm for building maintenance unless the organisation is confident of a return on financial outlay, for instance in terms of possible rental returns or resale value should the need to move on arise. It also promotes a tendency toward a more standard approach in that more specialised or idiosyncratic buildings are more likely to be hard to sell on or to do so quickly.

Many organisations, of course, are not motivated primarily, or at all, by profit: schools, universities, prisons and the National Health Service come to mind. The most effective use of their resources is still of importance, in some ways even more vitally so. The University of Oxford, for instance, is not likely to be able, just like that, to decide to withdraw from the provision of higher education of the highest quality. However, to enable it to keep itself in the vanguard of knowledge and research, to expand and continuously improve its provision, it needs to use its

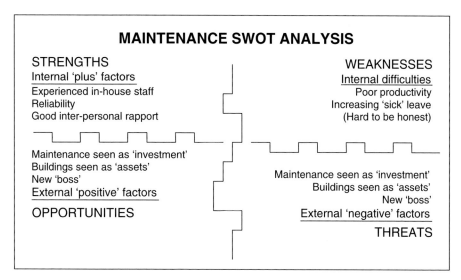

Figure 3.2 Example SWOT analysis.

buildings efficiently. While it could decide to develop an additional or 'satellite' facility in another place, whether that be in Birmingham, or Berlin, or Beijing, it is not very likely to 'do a bunk' and decamp from Oxford. It is thus compelled to make the most of what it has. The University is of course fortunate in that its heritage of wonderful architecture is known around the world; the 'dreaming spires' convey the image of the 'Oxford' education product. That same stock of buildings, however, also presents an organisational constraint and maintenance obligation. Whilst it may be acceptable, perhaps even desirable, for the buildings of the University of Oxford to display some of the patina of age, it will not do for them to look 'down at heel' and in desperate need of repair and maintenance. The condition of buildings and the facilities within are a tangible representation of the organisation, a visible manifestation of what it stands for and its priorities. The maintenance strategy must align with the organisation's strategy overall.

An organisation's strategy will typically be determined by its board of directors or governors or similar; the highest decision-making body in the organisation. The strategy presented to the board will be the product of possibly many person-days of effort by the organisation's principal officers. Sometimes a 'strategy day' may be organised as part of the development of the proposed strategy. This may be led by a trained facilitator, someone seen to be independent, with no personal axe to grind, no preconceived answers, or even predetermined questions perhaps. Approaches and techniques (fads?), such as appreciative enquiry, brainstorming, blue-skies thinking, buzzing and pyramiding, may be employed. It is important to maximise opportunity to think the impossible – therein may be the kernel of something that becomes the key to the way forward. For those of a more linear, reductionist or deterministic disposition much may be contributed through a fairly quick and rough-and-ready 'SWOT' analysis. Figure 3.2 shows an example SWOT analysis for a maintenance team where a new boss has just arrived (or is about to arrive).

SWOT captures initial thoughts on strengths, weaknesses, opportunities and threats, enabling discussion of and preliminary determination of where priorities and preferences for future directions may lie. There may be a need for a more formal options appraisal, perhaps based on a more thorough market analysis, maybe some market testing, and assessments of capability and risk. Where there are technical issues to be weighed, there should be firm recommendations presented for the benefit of non-technical board members.

What, where, when, how, (and how often), who and why

I do not intend to dwell here on the development of the organisation's overall corporate strategy, although the facilities manager may be part of the team that produced it. We are focused here on the development of the maintenance plan in the context of the organisation's strategy. The Chartered Institute of Building (1990) and Royal Institution of Chartered Surveyors (2000) have both issued guidance on the development of maintenance plans and strategy.

Why a maintenance plan?

To take the last question first, i.e. 'why', it will be clear that I think this generally the most important and challenging question. We have covered to an extent already why we should have a building maintenance plan; it is needed in order to give effect to the organisation's strategy. An inadequately maintained building will hamper the organisation in the fulfilment of its aims. A poorly maintained building may just about suffice to enable a set of operations to be carried out in it. However, if the operatives are not happy or if the quality of the product is impaired, there may be losses in productivity or in reputation that may in turn also reflect on the organisation's 'bottom line'.

It has been said 'fail to plan; plan to fail'. This is a way of saying how important it is to have a plan in place. It enables tracking to be done; a progress chart can be produced in the form of a graph or a bar chart or Gantt chart plotting work complete (or percentage done) against the amount planned to have been done by that time. Often quite complex flow charts or networks or precedence diagrams can be prepared, comparing actual against planned, and producing new plans in response to 'what if' scenarios, plotting when the works may be complete at the current rate of progress, or what needs to be done to remedy the situation. It is unusual for works to be ahead of schedule. I sometimes have a heretical thought when I think of how things might be if the resources spent in monitoring and recording progress were instead invested 'at the work-face' in actually *doing* the work. To not have a plan would be seen by many as quite unprofessional.

The act of preparing a plan makes it more likely that the work needed to repair and maintain the building will be identified and carried out. To prepare the plan, assessments are needed on a consistent basis, with all relevant elements considered. This is a professional activity. Although someone other than a construction-related professional may be able to process these assessments and to produce a maintenance plan, it is important that decisions on when a repair or replacement action should take place is based on firm technical assessments of the elements and their inter-relationships. For instance, while it may be possible to plan for replacements of roof slates 60 years after their fixing, it may be that their condition is significantly better or worse than may be expected. It will also make sense to carry out any repair work on chimneys at that same time, or to have them taken down if they are no longer required.

A possible downside risk of preparing a maintenance plan is that it may instil too short-term a focus. For instance it may focus on work needed to repair those defects which are readily apparent, especially if it is known that budgets are tight. Maybe the maintenance plan will militate against improvement because of a more immediate focus. There is a need to stand back and apply a longer-term perspective, 5–10 years being typical. It may be that data collected on this basis will suggest that it would not be worthwhile to carry out work to 'patch and mend' in the light of the longer-term needs. In this regard a view of the long-term contribution of the building to the organisation's strategic needs is invaluable.

What?

So what should a maintenance plan contain? What, therefore, should it be based on? It should contain all the work identified as needing, or recommended, to be done on the building(s) in question over the period to which it relates. It should set out what it does and does not contain, and any limitations on its applicability. Some maintenance activities planned for, say, 5–10 years hence may need to be tagged with a proviso that the condition of the elements concerned may deteriorate more or less quickly than projected; that funds may have been insufficient to have executed works needed to precede the activities in question. The long-term prognosis of the building's future may change over time, rendering the proposed activity no longer appropriate.

It may not be necessary to inspect every square inch (or millimetre) of every part of every building. A lot can be ascertained from judicious sampling. For instance, on a housing estate of 500 houses of identical design, or a tower block, it may be sufficient to survey a 10% sample. The location of the sample properties will need to be determined with intelligence, rather like choosing where to dig trial-pits or boreholes for a site investigation. If the samples show broadly similar conditions, then the site may be considered to be broadly consistent across it; if on the other hand there are significant differences, then more samples will be required to discern boundaries. For instance, when undertaking surveys to prepare a 10-year programme of repairs and upgradings for an outer London borough, such investigations identified significant differences in condition of houses of similar design. Further investigation identified that the builder of one part of the estate (that in less good condition) had gone into liquidation and that the remainder was completed by another contractor.

An experienced surveyor will be able to determine a great deal about a structure and its component parts from a visual, non-destructive inspection. Much more will be said about survey processes and defect analysis in Chapters 10–15. What is seen on the surface is a clue to, or sign of, what may be going on beneath. What is being observed is a symptom, to which intelligence needs to be applied based on context and experience. For instance, if dampness is detected in a building it may stem from one or more of many causes. Possible causes could be:

▪ water ingress, from perhaps a broken, leaking or blocked and overflowing pipe or gutter;
▪ a crack though which water is entering;
▪ rising damp;
▪ condensation;
▪ an unreported spillage of liquid.

Only when the actual cause has been determined can the requisite remedial action be called up. It may be possible to identify straight away a broken gutter immediately adjacent to the dampness, but that is not to say there may not be condensation also. If there is no readily apparent source, then further investigation will be required. It may make sense to continue with the survey and to put in hand a number of further investigations as another mini-project or perhaps to allow for these to be carried out as part of a package of remedial work. A provisional sum could be included in a contract pending determination of the remedial work to be done.

Some areas of work are specialised enough to warrant being done by someone with that specialist skill. Gas and electrical work, structural concerns, testing of drains and dealing with suspected asbestos products are good examples. Some installations require periodic testing, usually annually, and the provision of certification by licensed personnel or firms. No-one should contemplate surveying these areas without the requisite skill; not only will they stand the risk of missing important faults, but they will be professionally negligent too in going beyond their competence.

Where?

Surveys will need to take place on site. There is no substitute for seeing, and maybe touching, and sometimes smelling or listening to, the constituent parts of the building(s). For instance, surfaces may be rough and uneven; dampness and rots can smell; hollow plaster 'rings' when tapped. Preparatory work, for instance planning the survey, can be done in the office, as can follow-up research, and data from surveys can be transcribed or otherwise transferred to a database for analysis. Alternatively data transfer can be made direct from a handheld computer used for recording the data in the field.

What locations in the building(s) should be surveyed? All those likely to feature in the maintenance plan, no less and no more, using sampling as far as makes sense. This may involve going back to revisit parts of a building once a particular problem has been identified, recognising that similar faults may have been missed, in order to assess the extent of the problem. There tend to be patterns of failure repeated across similar locations. Areas that tend to give rise to problems are:

- where elements come together;
- unusual design details or materials used;
- exposed areas;
- voids of various kinds;
- buildings on sloping sites.

Places particularly worthy of inspection therefore include:

- eaves intersections;
- solid external walls (evidenced by brick headers);
- rendering;
- window heads, reveals and cills;
- floors at or below or close to ground level;
- intermediate timber floor/wall junctions;
- chimneys, due to their exposure, and flashings;
- slate roofs, due to their age.

Properties should be inspected methodically; externally starting at the top with roofs, chimneys and roof structures, such as lift motor rooms, and working down the building, and looking at all elevations. It may make sense to assess all external walls together; all windows together, and so on; or it may make sense to survey and record elevation by elevation if it is not easy to move around the exterior, for instance due to fencing and access problems. Similarly, internally there is a choice between surveying and recording room by room or by service, e.g. all electrical installations together; all sanitary appliances; water distribution; etc.

Figure 3.3 is an example of a pro forma data collection sheet used to capture information about the attributes and condition of the external elements of a stock of social housing. The surveyor fills in the appropriate slot on the form which is then read by an optical mark recognition machine and input directly to computer for storage, analysis and future reference.

When?

There are a number of issues here. When should buildings be surveyed; at what times of day, week, year; at what intervals? When should maintenance work be carried out?

ABERDABER DISTRICT COUNCIL HOUSING STOCK SURVEY
Attribute/Condition Data Sheet Date: .... Surveyor:...

Note/sketch/Photo Ref:
_____
_____ J
_____
_____ A
_____ B

Address/Label

UNIQUE REFERENCE

0	0	0	0	0	0	0	0	0	0
1	1	1	1	1	1	1	1	1	1
2	2	2	2	2	2	2	2	2	2
4	4	4	4	4	4	4	4	4	4
7	7	7	7	7	7	7	7	7	7

Category	Code	Option A	Option B	Option C	Option D	Option E	Condition
CHIMNEY Main	50	Ridge 0 / Brick 5	Slope 1 / Conc. 6	Flashed 2 / Render 7	Tall 3 / Metal 8	4/+ Pots 4 / Shared 9	0 2 4 6 8 / 1 3 5 7 9
Other	51	Ridge 0 / Brick 5	Slope 0 / Conc. 5	Flashed 2 / Render 7	Tall 3 / Metal 8	4/+ Pots 4 / Shared 9	0 2 4 6 8 / 1 3 5 7 9
ROOF	52	Flat 0 / Tiled Valy 5	Pitched 0 / Metal Valy 5	Bonnet Hip 2 / Flat Valy 7	½R/AngHip 3 / Step/Stag 8	Dormer 4 / EaveHtVar 9	
	53	Plain Tile 0 / Clay 5	Interlock 0 / Conc. 5	Profiled 2 / Slate 7	Felt 3 / Asb.Cem 8	Asphalte 4 / Metal 9	0 2 4 6 8 / 1 3 5 7 9
EAVES/ RWG	54	RtfrEnd 0 / Plastic 5	Fascia 0 / Cast Iron 5	Soffit 2 / Concrete 7	½Round 3 / Asb.Cem. 8	NoStdRWG 9	4 0 / 8 1 3 5 7 9
EXTL PLUMBING	55	SVP 0 / Plastic 5	Wastes 0 / Cast Iron 5	Combined 2 / Metal 7	Gully 3 / Asb.Cem. 8	Hopper 4 / Lead 9	0 2 4 6 8 / 1 3 5 7 9
BRICKWK	56	Solid 0 / Facings 5	Cavity 0 / Calc.Sil 5	Cav.Fill 2 / Commons 7	BrkQuoins 3 / BlackMrtr 8	Spall min 4 / Spall maj 9	0 2 4 6 8 / 1 3 5 7 9
OTHR WALL FINISHES	57	Blockwk. 0 / A-C Slate 5	Conc.Panel 0 / Metl.Sht 5	Full Rndr 2 / Timb.Bdg. 7	½Rndr 3 / Plas.Bdg 8	Tile 4 / Asb.Bdg 9	0 2 4 6 8 / 1 3 5 7 9
LINTELS	58	ConcStone 0 / UntrSteel 5	Flat Arch 0 / 5	Arch 2	Timber 3	GalvSteel 4 / NotVisibl 9	0 2 4 6 8 / 1 3 5 7 9
CILLS & SURROUNDS	59	Conc Stone 0 / Proj.Cill 5	Brick 0 / Drip 5	Brk.Specl 2 / Surround 7	2 CseTile 3 / W.Box 8	1 CseTile 4	0 2 4 6 8 / 1 3 5 7 9
WINDOWS	60	Casement 0 / Nitevent 5	H.Slide 0 / Subframe 5	V.Slide 2 / StoryFrm 7	Pivot 3 / Timber 8	Fanlight 4 / Proj.Cill 9	0 2 4 6 8 / 1 3 5 7 9
	61	Plastic 0	Alumin. 0	Galv.Stl. 0	UntrSteel 0	Mixed 4	
DOORS	62	Old Style	New Style 1	CombinFrm 2	No W/bd 3	3/+ 4	0 2 4 6 8 / 1 3 5 7 9
BAY/ Bayroof	63	Brick 0 / Tile Roof 5	Render 1 / Slate Rf 6	Tile/Bdg 2 / Conc.Rf. 7	2 Storey 3 / Metl.Rf. 8	UF Oriel 4 / Felt Rf. 9	0 2 4 6 8 / 1 3 5 7 9
CANOPY/ PORCH	64	Combined 0 / Tile Slate 5	Recess 1 / GRP 6	Pilr/Pier 2 / Brick 7	Posts 3 / Metal 8	Cantilv 4 / Felt 9	0 2 4 6 8 / 1 3 5 7 9
PRIVATE BALCONY	65	Integral 0 / BrkBalus 5	Cantlvr 1 / MetlBalus 6	Columns 2 / GlazBalus 7	Roofed 3 / ConcBalus 8	Walled 4 / InadDrain 9	0 2 4 6 8 / 1 3 5 7 9
EXTL DECOR	72	Doors 0 / Boarding 5	Windows 1 / Paint 6	Eaves 2 / Stain 7	Barge 3 / 8	Walls 4 / WallsReq 9	0 / 8 1 3 5 7 9
	74						
OUT- BUILDINGS	83	Attached 0 / Tile Slate 5	Paired 1 / SheetRF 6	Grouped 2 / FlatRf. 7	Brk.Wall 3 / Glazed 8	Conc.Wall 4 / RWG 9	0 2 4 6 8 / 1 3 5 7 9
PATHS & STEPS	87	Indiv. 0 / NoDrStep 5	Shared 1 / OthrSteps 6	Flags 2 / Ramp 7	Asphalte 3 / Handrail 8	Unmade 4 / Tunnel 9	0 2 4 6 8 / 1 3 5 7 9
PUBLIC BOUNDS	93	Ret.Wall 0 / Post/Wire 5	Timbr 1 / Chainlink 6	Painted 2 / MetlPosts 7	Hedge 3 / TmbrPosts 8	MultiBndy 4 / ConcPosts	
& GARDENS	94	Brick 0 / FrontBndy 5	Block 1 / SideBndy 6	Render 2 / RearBndy 7	Commons 3 / ComnlFrnt 8	InadCopng 4 / ComnlRear 9	0 2 4 6 8 / 1 3 5 7 9
GATES on PblcBnds	97	Sindle 0 / Metl 5	Double 1 / BrickTimbr 6	Tall 2 / MetlPosts 7	Two 3 / TmbrPosts 8	Tree 4 / ConcPosts 9	0 2 4 6 8 / 1 3 5 7 9
	98						0 2 4 6 8 / 1 3 5 7 9
	99						0 2 4 6 8 / 1 3 5 7 9

Figure 3.3 Example data collection pro forma.

Inspections should not be carried out particularly early or late in the day; not only may the light not be good, hindering the quality of observation, but residents and passers-by will be suspicious. It is good to notify occupants and neighbours of your intended visit(s) and to let the local police know. You should carry an official identification card and/or a letter of authority; these should be carried at all times, and shown when seeking entry and whenever challenged at any other time. Weekends are also not good times as colleagues may not be available to confirm your credentials. It is good to carry out surveys just after it has been raining; this enables defects in rainwater collection systems to be seen and related damp patches, while not inhibiting the recording of data (which is more difficult to do in the rain).

Where the survey extends to many buildings, it is prudent to limit the duration of the inspection to about 6 months. This is because data becomes increasingly out of date from the day it is collected, and it is important to try to base judgements of maintenance priorities on reasonably consistent data. It also makes sense to make the most of the longer periods of daylight in the summer months, and to avoid the short days of winter when surveyors are also prone to get cold and wet and to have more days when they may be unable to work.

It is more valuable for the surveyor to make an assessment of the urgency and type of repair or replacement work required than to merely record an element's condition. Together with an assessment of cost of the work this enables a maintenance plan to be prepared on a firm basis. It does not normally make sense to expect the surveyor to assess more accurately than to a year or two over a 5-year forward view. The survey form illustrated (Figure 3.3) seeks the assessment of action early, mid- or late programme, where early will normally mean 'Year 1' (but may become Year 2); mid Years 2 and 3; late, Years 4 and 5. More urgent work was coded 9; and areas requiring further investigation coded 8; 0 indicated no work required in the review period.

Sometimes a 10-year planning horizon may be sought; this makes some sense in terms of forward planning as long as it is borne in mind that projections will be increasingly uncertain the longer the forward look. It makes sense to carry out surveys at 5-year intervals with intermediate annual reviews to 'tweak' plans in response to actual progress and changing priorities. For instance, regulatory requirements and the economic climate will change from time to time, making it sensible to accelerate some parts of the plan and to reassess others.

The pattern of execution of maintenance work is also worthy of consideration. Does it, for instance, make sense to carry out a lot of electrical upgradings over a 2-year period, then none the next year, and then back to a full programme. This may be thought unimportant if planning to engage contractors rather than to use in-house staff, but it may have effects on tender prices and availability of skilled electricians. It may be less important if operatives are multi-skilled at the requisite level, but such tradespeople are unusual.

How?

How is the survey to be organised and executed? How is the maintenance plan to be prepared and evaluated?

Methodical planning is important and will repay in terms of efficiency. How long is the survey to take? This will depend greatly on the depth of detail to be collected, which in turn will relate to the purpose to which the data is to be put. If a substantial database is to be created, it is worth giving serious consideration to how much the data is likely to be used subsequently, and how often it will be updated. It is not worthwhile to collect and store data for which there is little or no likely value. It is worth carrying out a pilot study to assess the length of time needed to collect data, to refine the survey documents and processes and to assist in achieving consistency of understanding and assessment across the survey team, if there is one.

A quantity surveyor will be particularly useful when it comes to balancing priorities and preparing the proposed maintenance plan. Almost certainly the plan will need the approval of a manager or director or committee taking decisions about financial expenditure and the accuracy of the data and forecasts on which the plan has been based. Someone with recent experience of the preparation and execution of a maintenance plan will have credibility, whether they are acting as a consultant or are in-house. The facilities manager and maintenance manager will almost certainly be involved in the development of the plan and its approval, at least as a draft, before it goes forward to the board (or whoever is delegated the responsibility) for approval.

Communication

Again the what, when, how, etc. questions are pertinent. And why is communication important? Because we need to be able to advise or instruct others to do that which is needed; and because we are bad at it. There is much scope for misunderstanding, and, because potentially many people will be involved, there are many hundreds of communications likely to be made each day. There are many books on the art of communication, and pitfalls of miscommunication. There are, for instance over 3000 books on communication in the Oxford Brookes University library catalogue, with over 400 titles related to business communication alone. We will restrict ourselves here to matters relating to surveys and maintenance planning!

Some people are more assertive than others; some are firmer that a particular action must be taken, and immediately, while others may be more prepared to wait. We can but try to understand these differences and take them into account when taking decisions, and in implementing them. Reference was made in Chapter 2 to teamwork roles as explored and explained by Belbin (1981), for instance. Other authors speak of an individual's preferred learning style (e.g. Honey & Mumford, 1986) or personality type (e.g. Myers-Briggs, 1995) as being relevant to how an individual may prefer to receive information or respond better to it. Hofstede (2001) and Trompenaars and Hampden-Turner (2004) highlight the importance of understanding cultural differences. Some people like to be given crisp, direct instructions; others prefer to be politely requested, even talked around: 'I would be awfully pleased if you could perhaps . . .'. The right thing must be asked of the right person, at the right time and in the right way. As has been said, 'When in Rome, do as the Romans do'. For instance, I am given to understand that an on-road request for confirmation in Zimbabwe, 'Is this the way to Harare?' is likely to be answered in the affirmative, irrespective of whether or not that is the way; it is considered to be the polite and therefore correct answer. The better question is: 'Which way is Harare please?'

With face-to-face communication, there is plenty of scope for clarification of meaning through discussion, tone of voice, body language, taking time to check understanding, to question, express reservations, agree, accept, confirm, reinforce, decline or reject a request or instruction. Written communication, especially when formal, is often devoid of such nuances of expression; email even more so – indeed it can seem quite cold, terse and unfeeling. The immediacy of email is attractive and beneficial in many ways, but can be dangerous if time is not taken to think through the best response and to put oneself in the position of the recipient.

With maintenance planning, clarity is key – clarity of objective or purpose and clarity of expression. It is useful to develop, and apply, common standards and means of expression: for instance, the use of standard clauses and phraseology for the description of maintenance and repair tasks. The Chartered Institute of Building recommends a standard phraseology for condition reports (CIOB, 2004). The National Building Specification (NBS) and National Schedule of Rates (NSR) take this approach and are used extensively in relation to construction and repair specifications. Thus those drawing up and those implementing maintenance plans develop a common understanding and experience of what is meant and what is to be done. This reduces scope for doubt and expensive difference of opinion and interpretation. In any system where people are involved there is scope for misunderstanding; we can but try to reduce its occurrence and its effect. We must, however, also expect the unexpected, especially when dealing with existing buildings. Often the true nature of a problem is not fully revealed until work starts on the item *in situ*. Causes can be hidden from view and only seen when exposed by the removal of covering materials, panels, furniture or fittings. It is good to have some contingency.

It is important to develop and maintain a questioning mind and attitude. The dumbest question is the one you don't ask, for instance because you don't like to show your lack of knowledge. If unsure, it is much better to ask. This increases the likelihood of identifying the best solutions. While respecting knowledge and experience, nobody likes a 'know-all', especially when they are 'found out'; humility and respect for the inputs of others also help build team spirit. A good sense of humour (GSOH) also helps.

Case study: estate strategy in a UK university

The establishment is a fairly typical institution of higher education, in which teaching, learning and research are carried out across a diverse range of subjects, with associated administration and residential accommodation. It has to meet changing accommodation needs in the context of changing expectations and priorities.

Background

The institution had been created from a number of components – an art school, a technical college and a teacher training institution – located on two campuses a few miles apart. Nursing education was carried out in a number of further properties. Changes in the external context, including the business environment, brought three further major sites, with their very varied buildings, into the university's portfolio.

The estate has a total gross floor area of approximately 160 000 m^2, of which about 30% is residential (halls of residence). This comprises some 180 academic, administrative and residential buildings set within over 190 acres of grounds. Many of the buildings provide mixtures of functions within them. The largest contingent of buildings was constructed in the 1960s, with materials and design features typical of their day. Maintenance policies and structures were much as inherited from the constituent institutions with largely local authority/public sector-based procedures.

Issues

The condition and 'fit' of the estate to meet changing needs and expectations were perceived to be declining. Much of the information on the property stock was anecdotal – held in the memories of technical staff – and therefore inconsistent and insufficiently detailed or reliable for planning purposes. Such data as was held demonstrated a poor and deteriorating situation:

■ There was a maintenance backlog of over £30 000 000.
■ Utilisation of teaching rooms was as low as 20% (and only half that at unpopular times such as 5–6 p.m.).

- Support space for student learning was sparse and of poor quality.
- Research facilities were poor.
- Access, egress and accommodation for students and staff with disabilities were unsatisfactory.
- Residential accommodation was unfit for the conference market, limiting income-generating opportunities.
- Energy efficiency was of growing concern in the context of climate change and increasing energy bills.

In order to prepare realistic bids for significantly enhanced expenditure on the stock, it was necessary to assemble supportive documentary material. This demanded a refocusing of effort and consequential changes to the structure of the organisation and roles and responsibilities of people within the department.

Proposal

The Higher Education Funding Council for England (HEFCE) has an interest in the efficient operation of universities and a role in supporting their development. Before making substantial investments in such developments, including physical developments in the form of new or substantially improved buildings, it is both prudent and reasonable that they should be satisfied as to their long-term value. The development, presentation and acceptance of an estates strategy is a vital prerequisite of such physical developments. New and improved buildings must be justified.

Data was collected consistently across the estate, assembling records in such fields as:

- floor areas (nett internal area (NIA) and gross internal area (GIA) and space efficiency: NIA/GIA);
- space per student and per staff member (measured in full-time equivalents (FTE));
- utilisation;
- building condition and functional suitability;
- property costs (actual, annual and per square metre);
- costs to upgrade;
- energy costs.

A masterplan was to be prepared for the physical development of the three major campuses against which planning applications could be made and supported. Four local authorities are involved – the city council for one campus, two district councils (one for each of the other two campuses) and the county council. Significant new buildings are proposed for two of the campuses and a rolling programme of repairs and improvements for all of them. Major redevelopment of residential blocks has been carried out using the Private Finance Initiative (PFI), enabling a significant increase in both quantity and quality of provision.

Progress

The directorate reviews its effectiveness and reports annually to the board of governors its progress against the objectives set out in the strategy. Now 5 years into implementation the director has been able to report substantial progress. A traffic light system of reporting has been used: across nine agreed objectives, progress is assessed as 'green' on all but one – space measures – which is considered as 'green/amber' where the university's redevelopment plans are putting additional stress on current space utilisation.

After undertaking a detailed study of ongoing maintenance needs, a decision to eliminate a large chunk of old asbestos-lined building was taken. This gave the maintenance team the opportunity to make progress on the maintenance backlog while clearing the way for much needed and long overdue major capital works.

Significant achievements include:

- development of 10–15-year campus masterplans and residential strategy;
- completion of four major new buildings, including two specifically for research;
- refurbishments of teaching space;
- creation of two innovative 'student-centred learning zones';
- consolidation of academic schools on to single sites;
- major improvements in arrangements for disabled staff and students;
- 26% reduction in maintenance backlog.

A major restructuring of the management of estates- and facilities-related staff has been put in place. This has involved the creation of a dedicated in-house team for project management, specifically related to the masterplanning, briefing and development of the programme of new and refurbished buildings. This has included a substantial increase in staff resource and capability, creating a much more forward-looking focus. This also enables, it is hoped, an improved quality of build, value for money and programme certainty.

Assessment

Comparative performance as identified through the nationally collected and HEFCE-supported estate management statistics have demonstrated that the institution has:

- a space efficiency (NIA/GIA) of 0.8, which is good (upper quartile of higher education institutions (HEIs));
- utilisation of 0.38 (upper quartile, median = 0.25);
- buildings in Condition A ('as new') or B ('sound with only minor defects') improved from 57% in 2004/5 to 66% in 2006/7 (median = 71%);
- functional suitability improved from 44% (2004/5) to 55% (2006/7), but this is still in the lower quartile; median is 78%; upper quartile 86% – a measure of the need for and scale of improvement required, and planned for;
- property costs of just over £100/m^2 (median range);
- energy costs/m^2 of gross space of just over £11; median = £12.65. A major programme of energy-saving and carbon-reducing measures is in development.

The department is seen as a professional part of the university, key to its effective operation and transformation. Its profile has been significantly enhanced; it is no longer backward-focused, looking to a massive and increasing maintenance backlog. It has a positive present and future based on an ambitious programme of investment that is seen as central to the university's purpose and its ongoing development. The vocabulary is positive, featuring plans, programmes, improvements and implementation; and the directorate has recently achieved Investors in People status.

Summary

This chapter has discussed maintenance planning. It has outlined the development of a maintenance plan, devising a checklist for the process. A case study has also been presented to illustrate the process. The following chapter examines maintenance from the point of view of the client, who may be less concerned about processes and more concerned about the end product.

References

Ansoff, I. (1965) *Corporate Strategy: An Analytical Approach to Business Policy for Growth and Expansion*. McGraw-Hill, New York.

Belbin, R.M. (1981) *Management Teams: Why they Succeed or Fail*. Butterworth-Heinneman, Oxford.

Chartered Institute of Building (1990) *Maintenance Management*. CIOB, Ascot.

Chartered Institute of Building (2004) *Guidance Notes for House/Flat Condition Report Forms*. CIOB, Ascot.

Fowler, H.W., Le Mesurier, H.G. & McIntosh, E. (eds) (1934) *Concise Oxford Dictionary* (3rd edn). Clarendon Press, Oxford.

Hofstede, G. (2001) *Cultural Consequences: Comparing Values, Behaviours, Institutions and Organisations* (2nd edn). Sage, London.

Hollis, M. and Bright, K. (1999) Surveying the surveyors. *Structural Survey*, **17**(2), 65–73.

Honey, P. & Mumford, A. (1986) *Using Your Learning Styles* (2nd edn). Honey, Maidenhead.

Myers-Briggs, I. (1995) *Gifts Differing: Understanding Personality Type*. Davies-Black, Palo Alto, CA.

Ohmae, K. (1982) *The Mind of the Strategist: the Art of Japanese Management*. McGraw-Hill, New York.

Oxford English Dictionary Online (2008) http://dictionary.oed.com/cgi/findword?query-type=word &queryword=strategy. Accessed 19 November 2008.

Porter, M. (1980) *Competitive Strategy: Techniques for Analyzing Industries and Competitors*. Free Press, New York.

Royal Institution of Chartered Surveyors (2000) *Building Maintenance: Strategy, Planning and Procurement*. RICS Books, Coventry.

Trompenaars, F. & Hampden-Turner, C. (2004) *Managing People Across Cultures*. Capstone, Chichester.

4 The client

In the previous chapter maintenance planning was discussed, showing the development of a maintenance plan through the devising and application of a checklist for the process. A case study was also presented to illustrate the process. This chapter examines maintenance from the point of view of the client, who may be less concerned about processes and more concerned about the end-product. The material here is relevant to clients, to those advising them, and to those who need to understand clients so that they may better 'manage upwards'. We need to gain clients' approvals to maintenance plans and to maintain their continuing support in securing their effective execution. If we don't, we can expect a tiring and tiresome time.

The key decision maker

The client is the key decision maker. He, she, they or it pays the bills for the building and its running! This can create problems; it can also be seen as a great help, as making decisions and living with the consequences can be very difficult, and not everyone enjoys taking such responsibility. It can often be difficult to identify who is the real client. In a business run by a sole trader, this could be straightforward – the client is the proprietor. However, a sole trader is invariably working all hours just keeping the business running – winning contracts, doing the work and getting on to the next project. A proprietor with some, but little, knowledge of building-related matters may recognise the need for maintenance and the value of having this attended to professionally. He or she may know such a person or firm that will undertake or organise the maintenance work and engage them directly.

A small business may have expertise in-house, although this is unlikely in businesses other than those whose business is buildings-related. In a small partnership or business with a small board of directors or management group, it is likely that one of them will be allocated the responsibility for all buildings-related matters, including maintenance. This person will be, to all intents and purposes, the client, or the client's representative. Depending upon how the business is organised, that person may be able to make all decisions on building maintenance, or they may have limited scope for such decisions (for instance delegated powers to sanction expenditures up to a predetermined limit). They may need to report back or ask formally to a committee or subcommittee or other small group for approval. It is important to identify and understand these links and limits of responsibility, as there will be implications in how work is organised and sanctioned, and to allow for how long such processes may take.

A larger small organisation, especially one in which the quality and continuing efficient operation of the building(s) is important, may engage the services of a building-related professional, typically a chartered surveyor, to organise maintenance work on the client's behalf, as their agent.

The term managing agent may be used. Everything that the agent does is done in the name of, and on behalf of, the organisation that has engaged their services. The employing organisation puts their trust in the agent.

An agent will be expected to report periodically to someone in the organisation, to account for what they have done over the period of their engagement or over time since the previous report, or for some specific purpose. The agent may, for instance, wish to give or receive advice regarding the letting of a maintenance contract, or to seek instructions. The respective responsibilities and limits of action of the organisation and its agent will be determined by the contract between them (which may be an exchange of appropriately worded letters) and by the courts in the event of dispute.

Larger organisations are more likely to have an identified individual, or a small team, responsible for the identification of maintenance needs, and perhaps for the execution of some or all of the maintenance work. In this case the client, an internal client, may be another department in the organisation, or a committee or the board of directors who may determine priorities. Alternatively it may be that the identified individual or one of the team acts as the client's representative in engaging the services of, and overseeing the work of, consultants and/or contractors to undertake the planning and/or execution of some or all of the maintenance works.

As well as its size, the nature of the client organisation will also have an effect on how building maintenance work is organised. This is a gross generalisation but it is generally characterised that public sector organisations have layers of burdensome bureaucracy to be interfaced with and overcome, whereas the private sector is characterised as lean and efficient and therefore greatly to be preferred. This has been influenced and to a large extent determined by bad previous experiences of corrupt practices in the letting of large contracts in the 1960s. Local authorities and other public sector bodies dealing, for instance, with education, healthcare and housing projects and programmes have been seriously constrained by standing orders designed to eliminate unfair and dishonest practices and ensure public accountability.

In 1971, T. Dan Smith, the Leader of Newcastle City Council, and John Poulson, an architect, were sent to prison for corrupt practice in relation to the award of contracts for industrialised housing projects in the north of England. These bad experiences have inhibited innovation and advances in the procurement of building-related services, and this has sometimes been a brake on the ability of public clients to obtain best value for money. More will be said on this in Chapter 9. Public sector contracts will normally be let through a decision-making process in which several people of various backgrounds will be involved. In essence this group of people comprises the client in terms of that piece of decision making. It may be the same group that determined the procurement route, or a subgroup of it, or a differently constituted group. The execution of the work will almost certainly be entrusted to a construction-related professional or an appropriate firm acting as the organisation's agent.

Reminder: the client pays the bills

As may be said, 'he who pays the piper calls the tune'. It can be inconvenient, even undesirable, a nuisance, infuriating, and sometimes not the best technical solution, to do what the client determines. Clients will generally accept the advice and recommendations of their advisors, but of course they are not bound so to do and there will be instances where their knowledge and imperatives of their own business must over-ride other considerations. More will be said on this later in this chapter.

The building is a tangible asset. The client may own or let the building, and this is likely to affect how and to what extent they decide to maintain it. Buildings tend to decay over time from the moment of their construction. In fact some building materials and components will be deteriorating from the time they were quarried or left the factory. Standards sought by occupiers keep changing, generally increasing over time while the building's condition declines. Fortunately, across most of the UK and the developed world generally, property values maintain a general increase over time – they appreciate. Hence a building can become worth more even while it decays. But it may be worth more still if well maintained and kept up to date.

There is thus generally some incentive to maintain buildings to a good standard. However, in some parts of the UK, particularly in the former manufacturing and mining areas of the north of England, Scotland and Wales, there is depopulation associated with that de-industrialisation. Property values in those places are very much determined by the opportunities for regeneration. In some areas it may not make sense to invest in maintaining buildings which may have short useful lives, taking funds that might otherwise be needed to keep the business going. It is easier to justify expenditure on building maintenance when and where business is going well!

A well organised and maintained building may, however, make a valuable contribution to employer productivity. It is debatable whether increasing expenditure on maintenance (for instance improving a foyer area, or providing a better work environment) will so increase employer satisfaction and therefore productivity as to make such a substantial return on investment as to turn round a company, though it might. That would be a bold move. On the whole, the business response to difficult trading conditions and reduced profits is more likely to be to reduce staffing, the greatest cost. Building-related expenditure is, in fact, together with training budgets, the most likely to be cut.

Whoever in an organisation controls the budget is the person with most influence on the standard of maintenance of a building. A forward-looking chief executive officer (CEO), managing director (MD) or finance director (FD) will realise the potential contribution of buildings to the operational effectiveness, or otherwise, of the business. They will see to it that budgets for building maintenance reflect the needs of the business as represented in its strategic plans. A more utilitarian approach will see budgets increased in line with inflation, or inflation minus 1% (or 2%) or reduced (to make so-called efficiency gains) with little or no thought as to the effects. This approach is damaging to the physical quality of the work environment as manifested through, for instance, deferred redecoration programmes resulting in dull, dingy rooms with paint peeling, etc. It is also damaging to the sense of worth of the workers, including the maintenance team. It is vital for those involved with the creation and execution of maintenance plans to develop and maintain a good relationship with those who 'hold the purse strings' so that they can influence financial decisions.

'Money talks' and those who can 'talk money' will get a better hearing from the client than those who cannot or do not. Do not say 'extra cost' but do say 'better value' or 'investment' or 'pay-back period of less than 2 years'.

Wants and needs

'I desperately need that gold ring; I've got to have it now.' 'You can't possibly need another pair of socks; you have enough already, surely.' When I *want* something, I am likely to present it as a *need*. When someone else asks me for something, I am likely to question whether it is truly needed; whether it is needed now; could it wait and how long; would something cheaper do the job; and so on. There is a world of difference between wants and needs. This is not the place to debate the inequalities of affluence, world poverty and hunger, but even to mention such an issue points up the significance of appreciating the difference.

Anyone involved in building maintenance is likely to have to make choices, to prioritise one thing over another. This will be developed further in Chapter 6.

A study by Chinyio *et al.* (1998) of the needs of UK building clients, albeit in the context largely of newbuild projects, identified that the most frequently required needs were for functional buildings, satisfactory contract duration and building quality. Clients, and those acting on their behalf, are not good generally at determining priorities and turning them into briefs. Hansen and Vanegas (2003) have proposed the use of briefing automation to assist in improving design quality, and Gann *et al.* (2003) have assessed the value of design quality indicators as a tool. Smith (2005) emphasises the importance of cost budgeting, while Mbachu and Nkado (2006) stress the value of a conceptual framework for assessing client needs and satisfaction. It is important to determine client goals and priorities as there is significant variation between different clients (Lindahl & Ryd, 2007).

Clients should make such choices in relation to their buildings, their organisations and their people. For instance it is right and proper that the client should decide whether expenditure on foyer areas seen by their potential clients and customers is more important than expenditure on unseen back office areas used only by staff. The client should also set the standard in such terms as file storage – what should be kept, for how long, in what media, to what level of security, on/off-site, humidity controlled or not, and so on. What is wanted; what is needed? Professional advisors are available to advise if requested. The key word is *advise* – they make a report and offer recommendations for the client to consider and to accept or reject in whole or in part.

Such advice is likely to be sought in situations when one or more of the following pertains:

- The client has insufficient knowledge.
- The client has insufficient time to undertake the task him/herself.
- The advisor has recognised and relevant expertise.
- The client has worked with this consultant before, or has received good references.
- The client is in some distress and has been advised (perhaps strongly advised) to seek such advice.

As indicated, clients are at liberty to reject advice, although it is worth asking at that time why that is; it is wasteful, debilitating and potentially dangerous to ignore what should be good, sound and worthwhile advice. It is incumbent upon professionals to ensure that they understand the client's brief and better still to have assisted the client to ask the right questions. This is key. This is where wants and needs may be properly discerned and, if necessary, differentiated.

As a general rule, needs must be met, while wants may be seen as desirable higher or additional attributes or features. Thus, for instance, legislative requirements, such as the Building Regulations or Construction (Design and Management) Regulations (CDM) or the Disability Discrimination Act (DDA), must be complied with – that's a need; compliance is essential, or mandatory. Items that are not subject to regulation or control may be considered as desirable, i.e not essential, and these may be ranked in some way as to just how desirable they are. For instance it may be essential to clean kitchen surfaces several times a day; it may be desirable to clean office areas daily; it may be more important to be seen to clean foyer areas twice a day. Once the essential things have been attended to, the desirable matters can be addressed in an order of priority that reflects the organisation's strategic plans and priorities. A professional person can be expected to know the relevant legislative requirements and to ascertain and apply the correct criteria to give the client the best all-round performance in terms of his/her/its buildings and how they address the organisational need.

Legislation keeps changing, and increasing, both in quantity and in reach; more buildings and clients are included in the scope of regulation and the standards to be attained are becoming more onerous. It is not possible to give here a definitive list of applicable legal requirements; it would be very long and be out of date before publication. There is a number of suppliers of information services targeted at keeping clients and their advisors up to date. These are usually available as an annual subscription service, and although their costs may, at first sight, seem high, having up-to-date information immediately to hand is invaluable. Furthermore, its price compared with the cost of even trying to identify relevant legislation, let alone establishing its currency, is very low. Keeping up to date is time consuming and expensive; not being up to date is even more expensive.

The UK government has a policy of consulting on changes in legislation and its application; such consultations and the results thereof are available through www.info4local.gov.uk, which offers a twice-daily update via email alerts. Up-to-date Building Regulations and related Approved Documents are available at www.planningportal.gov.uk. A private company, Croners, has since 1948 provided a series of loose-leaf publications related to legislation relevant to a range of client organisations (www.croner.co.uk).

NBS Ltd. is a division of RIBA Enterprises Ltd., an offshoot of the Royal Institute of British Architects. It produces the National Building Specification, which provides standard wordings to describe building work of a good standard and incorporates advice on good working practices. It keeps up to date on such as Building Regulation requirements, British Standards and codes of practice of relevant trade associations (www.thenbs.com). The RIBA also offers an A–Z of other advisory organisations through www.ribaproductselector.com.

Some legislation makes building owners directly responsible for what goes on in their buildings and for any matter consequential upon the building. They have responsibilities to and for their employees and building users generally, and passers-by. For instance, the CDM regulations, referred to above, make the building owner responsible for the health and safety of all the construction workers involved in a building-related project and for those affected by the works. The Health and Safety Executive maintains an excellent website (www.hse.gov.uk).

How to determine and differentiate wants and needs

There will be competing and sometimes conflicting demands on the client. There are rarely sufficient funds to meet the wishes of all stakeholders and decisions need to be taken on what should be done. How to decide? This section is aimed at trying to assist in that process.

There may be times when a strategic direction can be determined; and other times when a comparatively straightforward yes or no (or not yet) decision is required. For some decisions, and some decision makers, it is important, perhaps imperative, to have all the information and/or to consider all the options. For others a firm recommendation from an appropriate, probably professional or expert, person is preferred; another may be more comfortable and confident making a choice between two or three possibilities. Many professionals will feel that the best client is one who delegates to them all the decision making, perhaps within an agreed budget.

I set out here a rather linear or deterministic decision-making process; those who are more divergent or ruminative may see this as rather restrictive – it could perhaps be used as a 'point of departure', a stimulus for asking at least some pertinent questions. That is the starting point for the process as set out here: to determine the right questions to ask, and hopefully answer.

Figure 4.1 shows in a simplified way that wants and needs should be differentiated and a balanced view formed.

The key order of priorities is:

- health and safety;
- wind-and-weathertightness of the building;

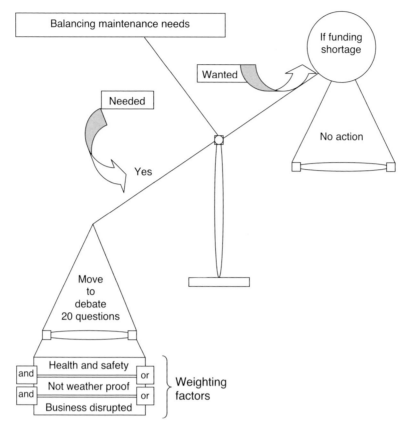

Figure 4.1 Differentiating between and balancing building maintenance needs.

▪ continuity of business operations;
▪ comfort of occupants;
▪ increasing efficiency, effectiveness and economy of operations.

The order of the last three items above may be debated and different organisations may order them differently. It can be a worthwhile exercise to have that discussion.

I offer below a '20 questions' approach to decision making:

1. What are the issues here?
2. What is the key issue?
3. Is it important? What is its significance? How important is this? Essential? Why?
4. Is it a legal requirement?
5. How pressing? Urgent? Is something already non-compliant? – in which case immediate attention is required, which may entail closing down temporarily some operation or part of a building.
6. Is the health and/or safety of anyone endangered or compromised?
7. What is the scale (magnitude) of difference this decision may make?
8. How high does the issue figure in the organisation's strategic business priorities?
9. What is the potential downside of not making (or delaying) a decision?

10. How satisfactory is the *status quo*? (the 'no change' option)
11. What history is there of requests/demands/responses in this area?
12. How does the issue relate to what else may be going on?
13. What knock-on effects might there be to additional, related or rival requests?
14. What is the potential downside of making a wrong decision?
15. Are you inclined toward approval in order to be liked, to avoid conflict?
16. If you are considering cost, stringency or availability, you are not seeing a need.
17. What evidence is available to support the request; and the decision?
18. Is it worth obtaining further information/a second opinion?
19. Are you fundamentally a procrastinator?
20. If the issue is still in question at this stage, it doesn't feel like a real need.

Practical application: example scenario

If it sounds easy to make decisions on what proposals to support, or, if 'the buck stops with you', to accept, perhaps it will be worthwhile to consider a scenario such as this. You are the head of operations in an organisation that has been recently taken over by a former competitor that is based in a city in another part of the country. The CEO has issued a directive that all plans are to be reviewed with the objective of improving profitability by taking out cost, and that in the meantime no non-essential expenditure will be sanctioned. The maintenance manager informs you that she received another complaint last week from an administrative assistant in the warehouse about the slow clearance of the w.c. when it was flushed.

1. What are the issues here?

There is quite a lot going on:

- Your organisation has been taken over.
- This is a recent (and major) change.
- The (new to you) CEO has issued a directive.
- Your job is at risk; you may be 'surplus to requirements'.
- The maintenance manager may feel similarly vulnerable.
- Similarly the administrative assistant . . .
- . . . and others who may be encouraging or discouraging the complaint, and the complainant.
- It's *another* complaint. How many have there been? About this issue?
- How slow a clearance?
- Has it been investigated? By whom? When? With what result?
- Could it be a water supply problem? A drainage problem?
- Could there be knock on effects, e.g. if sewage is leaking from the foul drain.
- What alternative arrangements are available/have been put in hand? Is this the only toilet in the building? Are others affected? Are there other toilets available?
- Why are you being told?
- Why now?
- Who and what do you fear? The CEO; the maintenance manager; the complainant; the staff generally; your spouse or partner? Making a wrong decision that you fear may cost you your job? Feeling disempowered? Losing face? Being indecisive?
- Re-runs of previous unsatisfactory situations?

2. What is the key issue?

This is hard to determine with any certainty. There are several technical issues. However there is a substantial 'human' element too. The cost to the organisation of loss of production or reduced productivity due to staff demotivation is potentially substantial; much higher than the likely cost of remedial work.

3. Is it important?

What is its significance? How important is this? Essential? Why?
For the reason stated above, i.e. the loss of productivity due to demotivation, it is important to be 'on the case', and to be seen to be so. If staff gain the impression that 'the (new) management' doesn't care about them, they will soon switch off. For as long as an organisation is employing people it is essential to keep them actively engaged in productive work; to not do so is to incur expenditure without related income. It is important to value staff as fellow human beings.

4. Is it a legal requirement?

It is essential to provide toilet and hand-washing facilities for staff; and it is generally desirable to do so for visitors too. Many shops for instance do not provide toilets for customers (although the UK government is consulting on demanding that such provision be made available to the public as a whole with the demise of public conveniences). Regulations stipulate the minimum provision in relation to occupancy and gender; they are not definitive however in terms of location. It may not be a requirement that this particular toilet be kept operational. It will however be essential that there is not an ongoing unsanitary situation – that would be a matter of public health concern. Some further on-site investigation will almost certainly be required.

5. How pressing? Urgent?

Is something already non-compliant? In which case immediate attention is required, which may entail closing down temporarily some operation or part of a building.
This consideration follows on from the previous one. Expecting staff to use a toilet in another building will convey a message of a kind; this may be unhelpful at this time of uncertainty. Closing down the warehouse, however temporarily it is said to be, would not only have a serious effect on operations, but may also be read by staff as a sign of things to come – a permanent closure. Prompt and considerate action is required at this critical time.

6. Is the health and/or safety of anyone endangered or compromised?

Health may be endangered here – publicly through leaking drains and privately if people are unable to use hand-washing facilities. Safety may be at risk if staff need to travel outside to another building, for instance in icy conditions.

7. What is the scale (magnitude) of difference this decision may make?

As indicated under 5, above, staff may read the situation (in)correctly. There is potentially a lot at stake here.

8. How high does the issue figure in the organisation's strategic business priorities?

It is, unfortunately, quite likely that the CEO, and his/her board and advisors, will not have worked through all the ramifications of the takeover. Even when there is a clear decision on the new shape of the organisation, what buildings are to be retained and for what purposes, and which staff are to be retained, there will still be detailed plans to be developed and implemented to get from current arrangements to the new state. Many transitional arrangements will be necessary. It may be that the CEO neglects to indicate that the well-being of his/her staff is a (or the) top business priority.

9. What is the potential downside of not making (or delaying) a decision?

Indecision, dithering and delay will gain nothing here; they will give the impression (and worse still, an accurate one) that staff, their concerns and welfare, are to be ignored, disregarded or dismissed.

10. How satisfactory is the *status quo*? (the 'no change' option)

What is the *status quo* here? If it were not for the takeover, there would have been procedures in place for dealing with this request before it became a complaint. This *status quo ante*, that is the position pertaining previously, would, presumably, have dealt with the situation satisfactorily – so why not stick with those procedures? It is doubtful that the CEO wants all such expenditures forwarded to him/her for approval; he or she would not be able to get on with the strategic planning. Perhaps it would make sense to agree a financial limit above which items would be referred for decision. Perhaps the CEO would delegate (or entrust) such decisions to you? Maybe you could (re)empower your maintenance manager similarly?

11. What history is there of requests/demands/responses in this area?

Would your maintenance manager normally (previously) have referred this to you? Has this toilet (or this member of staff) been a problem before? Has it been a problem often? In which case perhaps inappropriate materials (e.g. sanitary towels) may be being disposed of down the toilet? Have other toilets been problematic? Where do cleaners dispose of their materials? Does the warehouse give rise to a lot of complaints about maintenance; about issues generally? Have requests and/or complaints gone up (significantly?) since the takeover was rumoured/announced? The slow-draining w.c. will still need to dealt with, and soon, but it does give rise to these further questions.

12. How does the issue relate to what else may be going on?

In this case we know quite a lot about what else is going on. However, there may still be other issues that may be related, or unrelated. It may be that a relationship has started (or finished) between members of staff; that there are inter-personal rivalries, disagreements, private agendas and points to prove, alliances to be made. It is quite likely in this situation that staff will need reassurances and 'stroking' – that is, to be seen to be thought important and cared for. This must be attended to, and soon. A building that has been well looked after will be more marketable too, should the need arise from the strategic review.

13. What knock-on effects might there be to additional, related or rival requests?

The CEO and his/her advisors may have hoped that money would be saved through imposition of the directive. However, answers to preceding questions have suggested that the contrary may in fact be the case. To decline to deal with the complaint may result in several undesirable consequences. Acceding to the request may give rise to an increase in such requests, and consequential expenditures, but the goodwill engendered (rather than lost) may be considerable. This would be a good knock-on effect. The better condition of the building will be another beneficial effect.

14. What is the potential downside of making a wrong decision?

Loss of goodwill would be expensive. It takes a long time to build up trust and reputation, and little time to destroy it. Good news may travel fast; bad news travels faster. When people are upset or annoyed they tend to tell all their friends and acquaintances; they may contact the local newspaper. The national (and international) papers and broadcasting media may become interested. Your peers and competitors may hear of your plight and enjoy (and enhance) your discomfort; they may gain business at your expense. You may incur professional criticism and censure. You may be dismissed from your job, even if unfairly. You may become involved in litigation; you may find yourself in court and perhaps in prison if you are found to have been culpably negligent. You can be in as much trouble for what you don't do as for what you do do. These are good reasons for doing the right thing.

15. Are you inclined toward approval in order to be liked, to avoid conflict?

Most people like to be liked; few would set out to be disliked. There can however be a suggestion that someone is liked because they failed to make the hard decisions and to follow them through. 'Good' or 'effective' managers are those who sack people; the more that are made redundant, or laid off or let go, the 'better' the manager, in the opinion of some; it is almost a 'badge of office'. It is more important in business terms to be respected than to be liked, and, of course, to make the right decisions. People will understand, and respect however grudgingly, hard and unpalatable decisions when faced with reasoned and well presented argument. Expect disagreement though, and conflict, hopefully not violent though. Be prepared. Expect people to be upset, angry, tearful, emotional when faced with bad news. No amount of 'spin' or gloss can cover over bad news. In this case, being liked will be the outcome from considering the items above, but don't expect gratitude. You are only doing what you should and what staff see as fair and what they expect. Don't expect the CEO to be pleased either, though he may be, especially if he realises the potential mess from which you have saved him or her. He/she probably will not thank you for telling them that though. Conflict can be good. If this complaint had not reached you, who knows how matters may have developed? Complaints are good, although you may not feel good about receiving them and dealing with them.

16. If you are considering cost, stringency or availability, you are not seeing a need

The job has just got to be done.

17. What evidence is available to support the request; and the decision?

It may be of interest to know how often such a problem has been reported, what the responses have been and to what effect. If major work has already been carried out in relation to this slow-clearing w.c. that may help reinforce the unsatisfactory nature of the building, or at least that aspect of it. It will not however reduce the need to rectify the currently unacceptable situation. Considerations of the long-term future of the building may however suggest that making good (work related to and around the repair) may be less than otherwise would be warranted – for instance not redecorating, or resurfacing in cement or concrete without a further surface finish of brick or tile or carpet. The context may be relevant to the request; it is relevant to the decision.

18. Is it worth obtaining further information/a second opinion?

In this situation there is little to be gained by seeking further information; indeed delay may result in a worsening state of affairs, with this w.c., and perhaps others, overflowing with consequential increase in the hygiene problem to be cleared up, and a loss of motivation in the staff, who will feel ignored and devalued. A speedy and considerate response is required, undiminished by the longer-term property context.

19. Are you fundamentally a procrastinator?

In my own case, I don't like to rush at decisions – I like to get an overview. But I am not then concerned to have a lot of detail; I am fairly intuitive – I have a feel for what the issue is. That may seem to some like a prompt response; to others, procrastination.
Try to get to know yourself, how you work best and how your colleagues respond.

20. If the issue is still in question at this stage, it doesn't feel like a real need

Enough said, I say.

Seeking advice

There is an old saying in the legal profession that a person who represents himself has a fool for a client. This a way of saying that it is appropriate to seek and take advice from people who have good knowledge of a situation and the issues around it, and can refer to similar situations elsewhere and previously, and their outcomes. They are also not only in a position to give advice without bias or self-interest, but are also obliged as professional people so to do. That advice costs money, but it saves the client the time he or she may have spent researching and evaluating evidence and hopefully will be worth it by way of further time and costs saved that would have resulted from taking wrong or less good actions. This addresses the question of *why* to seek advice.

When to seek advice?

This includes *whether* to seek advice, and *how often* and *how soon*? It is important to seek advice in situations when any of the following may apply:

- You are conscious that you are 'out of your depth', possibly dealing with a big issue for the first time.
- Your spouse suggests you should.
- The possible down-sides of a wrong decision may be catastrophic or expensive.
- You would like the comfort of a well informed opinion or second opinion.

- When there are multiple alternative, possibly conflicting, options ahead.
- A business colleague, friend or relative may be seriously affected by your decision.
- You see danger ahead and have the time in which to seek advice.
- There are legal implications – e.g. contracts, employment (especially termination).
- You receive communications from, or intimations of a resort to, a solicitor.
- You seem to be spending inordinate lengths of time on apparently small and inconsequential, and repeating issues.
- You are suffering periods of ill-health, lethargy or tiredness.
- You are genuinely undecided and cannot 'toss a coin' to decide.

How often to seek advice

One should not seek advice too often – not only is advice expensive (you will usually pay more per hour or day than you earn), but your position will look weak if you appear indecisive, always seeking advice. Are you looking for someone else to blame if and when things do not turn out as anticipated (as will be the case on occasion)? You will need to be selective in when to seek advice. If you have recently obtained advice on a similar situation you may be able to extrapolate from one case to the other. However, it as well to keep in mind that the law and legal precedent, in addition to people's expectations, develop and change over time; we must keep up to date. The kinds of service referred to earlier, where for instance legal updates are issued monthly or quarterly on subscription and searchable archives are available (e.g. www.butterworths.co.uk, www.croner.co.uk, www.infolaw.co.uk) can be very useful.

How soon to seek advice

One should seek advice as soon as is reasonably practicable. There may be situations where your inactivity or slowness to respond may be taken to indicate a lack of concern – that may cost you loss of goodwill, or worse should you find yourself in court as a result of some disaster that came to pass because of your inattention. Not only may 'a stitch in time save nine', keeping costs and consequence in check, but an informal discussion and receipt of a wise word may enable the whole matter to be 'nipped in the bud'.

From whom to seek advice

It is not a good idea to seek advice from a friend. It may seem a low-cost way ahead; it may cost you a pint of beer or glass of wine or a meal out. It will cost you a lot more if your friend tells you what he or she thinks you want to hear; or if you ignore that advice however well informed because you valued it lowly. You may fall out as a result of acting upon (or not acting upon) the advice; that's a high price to pay.

It may be a good idea to seek advice from someone of whom you have good reports; that will give you some confidence. However you need to be sure that the matter on which you seek advice is one in which they are experienced and may specialise, even excel. You want the best advice, and it may take an expert less time to deal with your concern than for someone for whom it is a new or less well known area. The hourly charge of the expert may be higher, perhaps appreciably so, but the total charge may not be much different, and the advice should be better.

In circumstances where you are in dispute and your adversary has access to substantial funds it is important to obtain the best advice you can and you will have to consider the possibility of settling quickly, without argument and hopefully without acrimony. You may not be able to afford to win a protracted argument.

Be sure to consider what redress you may wish to have recourse to in the event of poor or inadequate service, and ensure that either there is appropriate indemnity cover or that you are prepared to deal with the possible consequences, however unlikely they may seem.

Giving advice

There will be times when you are the person who has the experience or expertise to be able to advise another or others. Be conscious that advice is worth what someone will pay for it – so be wary about giving free advice. A professional person may be liable for the 'advice' they give, even without charge, if the person acting upon it had reason to believe it to be accurate and reliable. This is an awesome responsibility, so you must be careful.

For instance, if you make recommendations for action you are going further than if you just observe and report upon a matter, such as in a property condition report. When carrying out a building survey, you must be diligent to record what you observe, and to observe what may be held to be reasonable that you should have observed. You also can be expected to advise your client and intended recipients of your report of situations that are dangerous or unsatisfactory, and the consequences of what you observed if no action is taken. Because it is difficult to limit your liability (Unfair Contract Terms Act 1977), it is important that your charges reflect that.

You must be utterly professional in your giving of advice. Do not work for friends. Do make sure you have a written agreement to the terms of your engagement before commencement. If you are being asked to give advice in-house, be sure to obtain in writing clear terms of reference, and make appropriate caveats relating to the limits of intended application of your advice. It is worth developing standard letters, forms of words and phraseologies and terms of engagement.

Taking advice

Why keep a dog and bark yourself? If you seek advice you should accept it and expect to act accordingly. The best advice may not be what you want to hear; that is not good enough reason to reject that advice. You may not be able to afford to act upon the advice; can you afford not to? You may need to reorder some priorities. You may be able to gain clarification on the acceptability of some delay or phasing of a response.

Once it becomes known that you have received advice as to a course of action, you will be in a difficult position should it also become known that you ignored or rejected it. If you don't expect to act upon advice, you may be as well to not seek it, although it is unlikely that ignorance or avoidance will provide much defence!

In addition or as an alternative to advice from an expert, it is also of course possible to seek advice from published sources: books, journals, legislation, papers of various kinds, in hard copy and online. Extensive databases of information and search engines, such as Google, Yahoo and Ask Jeeves, enable huge quantities of material to be accessed. There is a skill in developing methods for 'mining' the data to obtain relevant material, and you will still need to be able to evaluate it against the context in which you are working. In the latter part of the twentieth century there was considerable interest shown in the development of decision support (or intelligent) systems. These were intended to enable the determination of the most appropriate actions in relation to any particular combination of circumstances – the 'what if' scenario. Of course the answer very much depends on the algorithm on which the system is built. This could be based on a very substantial database of 'what did happen when ...', but it is also limited by that. It is unusual for a so-called intelligent system to act in an intelligent way such as a human might, truly learning from past situations and applying this selectively and appropriately (not unthinkingly or automatically, or robotically) to a new, always at least slightly different situation.

The situation into which advice is to be applied should be outlined at the outset; advice cannot be applied in a vacuum, unrelated to the context. Expect to act upon the advice, and record that you have done so.

Summary

This chapter has considered the nature of the client, and various aspects of clients. The client is affirmed as the key decision maker. He, she or it owns or leases or rents the building (or part of it) and wishes to see their money used wisely. It is therefore important to be able to distinguish and differentiate between wants and needs, to determine priorities and advise appropriately. The next chapter considers expectations that people may have of buildings and their maintenance, and that following discusses ways of prioritising.

References

Chinyio, E.A., Olomolaiye, P.O. & Corbett, P. (1998) An evaluation of the project needs of UK building clients. *International Journal of Project Management*, 16(6), 385–391.

Gann, D., Salter, A. & Whyte, J. (2003) Design Quality Indicators as a tool for thinking. *Building Research and Information*, 31(5), 318–333.

Hansen, K.L. & Vanegas, J. (2003) Improving design quality through briefing automation. *Building Research and Information*, 31(5), 379–386.

Lindahl, G. & Ryd, N. (2007) Clients' goals and the construction project management process. *Facilities*, 25(3/4), 147–156.

Mbachu, J. & Nkado, R. (2006) Conceptual framework for assessment of client needs and satisfaction in the building development process. *Construction Management and Economics*, 24(1), 31–44.

Smith, J. (2005) Cost budgeting in conservation management plans for heritage buildings. *Structural Survey*, 23(2), 101–110.

Useful websites

www.butterworths.co.uk. Accessed 16 February 2009.

www.croner.co.uk. Accessed 16 February 2009.

www.hse.gov.uk. Accessed 16 February 2009.

www.infolaw.co.uk. Accessed 16 February 2009.

www.info4local.gov.uk. Accessed 16 February 2009.

www.planningportal.gov.uk. Accessed 16 February 2009.

www.ribaproductselector.com. Accessed 16 February 2009.

www.thenbs.com. Accessed 16 February 2009.

5 Expectations

The preceding chapters have considered issues related to maintenance planning and how design may affect that. The nature of the client, or various aspects of clients, has also been reviewed. This chapter discusses what may be realistic present and possible future expectations that people may have of buildings and their maintenance. What are the matters to be considered; how are the required functions to be delivered; how may they be measured? What may the future hold? How can our buildings make provision for the future; can they be future-proofed?

Functionality

Our buildings must be able to continue to accommodate satisfactorily the activities they are there to support. That is the prime function of a building and the function of building maintenance is to support that. This linking of building maintenance to building functionality was established in British Standard BS 4778 in Part 3, Section 3.2, issued in 1991. That standard defined maintenance in relation to function:

> *the combination of all technical and administrative actions, including supervisory actions, intended to retain an item in, or return it to, a state in which it can perform a required function.*

Leaving aside for the time being issues related to the minutiae of meaning of individual words in that British Standard definition, it is clear that the authoring committee deemed it important to link the purpose and outcome of maintenance to something measurable – that the item 'can perform a required function'.

So what functions are required to be performed in or by a building? It will differ from one building to another. For instance, a town hall building needs to perform the functions of a town hall; a bank building those of a bank, whatever they are. What is expected of one town hall may differ from that of another; and it may change over time. Let us develop this example.

Across the United Kingdom, many town halls were built in the Victorian era to provide administrative centres for the new industrial towns and cities. These municipalities with their council members and officers to advise them were needed to initiate, control and maintain the new sanitary and other engineering services – roads, street lighting, drains, sewers, etc. They also needed facilities to collect the rates to pay for these services, and the staff (local government officers as we would now refer to them). These were generally housed together in a new building in a central location, with great competition between adjacent boroughs to have the grandest representation of their importance and wealth. The cities of the north of England obtained particularly fine new town hall buildings. Manchester, Sheffield, Leeds, Bradford and Bolton have especially magnificent examples.

However, over the years, the functions of local government have grown and changed and town halls have been extended and altered and added to in response. The ways that local authorities relate to those they are there to serve have also changed. For instance, district councils have had housing responsibilities added (and in some cases removed again to housing associations and other providers). Unitary authorities (those large enough to warrant having responsibility for all local government functions in their area) are also responsible for primary and secondary education (but not now further education or higher education) and social services. In many cases councils have sought to reorganise, repackage and decentralise their services to local offices and/or to outsource (or subcontract) services to other organisations to provide a service or services from other premises on the council's behalf.

Thus the whole nature of what the town hall is, and what function(s) it is there to provide have changed, and will continue to change. As services are moved from one council to another through periodic local government reorganisations, or expand and contract with changing demographics (e.g. fewer children, more elderly and frail), so buildings need to adapt and change. New departmental combinations, divisions and reconfigurations may need to be put in place, for instance the co-location of all children's services formerly separated in education and social services. Perhaps this may entail the construction of a new 'town hall extension' or 'annexe'. More likely, this may mean from time to time the acquisition or leasing of further accommodation, moving and disposal or creative reuse of current premises no longer required. Such moves may correspond with the coming to an end of a lease, in which case the local authority may have little interest in the future of a building which is it intending to vacate. However, many leases place full repairing obligations on the lessee; it will then be incumbent upon them to return the building to the lessor in a satisfactory condition – this can be an area of much dispute between surveyors as to the acceptable standard of the property.

In circumstances where the lessee is responsible for the maintenance of a building, it is important to develop a proper understanding of what that maintenance obligation is, and how it is intended to be met. This begs the question, perhaps thought unnecessary to question, 'what is maintenance'? The author was involved a few years ago with briefing surveyors for a government department on the subject of building maintenance. As we 'unpacked' the definition of maintenance in BS 4778, Part 3, Section 3.2: 1991 it became apparent that maintenance included actions *intended* (my italics) to retain an item in, or restore it to a (particular) state. That is to say that it does not mean that the actions need to have been effective for them to be considered to have been *maintenance*. By this definition, maintenance is an input and an intention, not an output or outcome. Thus it is insufficient to say that a building or part or service should be maintained, but that it should be well maintained, or, better still, to be maintained to a particular, defined and measurable standard.

Standards

The word standard can mean several things.

A Roman, or Imperial, or Royal, standard was something to be held aloft to rally the troops, to signify the presence of the person whose standard or principle or rule was to be upheld, sustained, fought for.

The Oxford English Dictionary (http://dictionary.oed.com) offers 30 definitions of 'standard' as a noun, and a further six as an adjective. Leaving on one side those that refer to botanical, ornithological, historical and obsolete meanings, those pertinent to our considerations are:

- *The authorized exemplar of a unit of measure or weight; e.g. a measuring rod of unit length . . .*
- *An authoritative or recognized exemplar of correctness, perfection, or some definite degree of any quality.*
- *A rule, principle, or means of judgement or estimation; a criterion, measure.*
- *A definite level of excellence, attainment, wealth, or the like, or a definite degree of any quality, viewed as a prescribed object of endeavour or as a measure of what is adequate for some purpose.*
- *Serving or fitted to serve as a standard of comparison or judgement (adjective).*

Perfection is not possible this side of heaven. The word 'perfect' means in one sense 'past'; we speak of something having happened as being in the perfect tense – it's over; it's done; it is finished; complete. There may be value in striving for perfection, but there are increasing costs and decreasing returns as perfection is approached. The higher the standard, the more that will fail to attain it – that is a cost and a waste.

Womack and Jones (1996, 2003, 2005a, 2005b) have written of seven wastes (*muda* in Japanese), popularised in *The Machine that Changed the World* (Womack *et al.*, 1990):

- overproduction;
- transportation;
- waiting;
- inventory;
- motion;
- overprocessing;
- defects and rework.

How might these wastes relate to maintenance; and might there be other wastes?

Overproduction

It is difficult to think of too much maintenance being done. However, there is a sense in which this may be the case when, for instance, all the fluorescent light tubes or windows in a building or on an estate are replaced when some are in good condition or have been already recently replaced. However, a kind of visual decay occurs with piecemeal replacements, especially when the new is to a different design from the original.

Transportation

This is a significant area of potential waste in building maintenance. There are few maintenance operatives or managers whose working day or week will be contained within one building; there will be travelling between as well as within buildings to inspect, organise and carry out maintenance and maintenance-related work. There will be transportation of materials to where they are needed, from builder's merchants and other suppliers to and within the buildings.

There will be waste material from the original, now inadequate, construction, and from over-ordering of the new material. There will be equipment to move to and from the maintenance work – ladders, tools, perhaps heavier lifting and moving equipment. The effective coordination and minimisation of transportation – logistics – can be time consuming but a worthwhile management task; and as the fuel used in transportation is generally from a non-renewable resource, oil, it is a very 'green' matter too.

Waiting

This too can be significant in building maintenance. How long is any particular maintenance job going to take? Who knows? Can it be worked out? How; with any degree of certainty? How long will it take to travel from one maintenance job to the next? Building maintenance often comprises many small jobs, often disparate in nature and to be carried out in a variety of locations. The time allowed to complete one maintenance task and get to the next may be too short. Then anyone required to attend for the second task (for instance to allow the maintenance operative to enter the building or the relevant part of the building which may be security-controlled) will be having their time wasted while waiting. On the other hand, if too much time or 'slack' or 'float' time is allowed, then the maintenance operative may be wasting time waiting for the other person to attend at the appointed time. Travel times can also be very variable depending upon time of day, traffic on overcrowded roads, parking problems, etc. Communications systems and devices such as the mobile telephone and palmtop computers have helped a great deal in being able to adjust timings on the day, not always conveniently for all however. Such devices can enable some people to act without due consideration of the value of other people's time, making changes at the last minute, perhaps in the extreme putting off work to another day and upsetting other people's programmes. Recording and analysing times taken to get to and to complete particular maintenance tasks can, and should, usefully inform future planning.

Inventory

While this may not seem as significant an area of waste as once it may have been, it is still significant. The days of large stock rooms full of spares or parts (inventory) are now long past. They have been largely superseded by systems of 'just in time' ordering and supply. Sometimes a replacement part may be made just at the time it is requested. If that is just following the time that the original part has just failed, then there is a sense in which the replacement is 'not quite in time'. Sometimes failed items will be a part of a component comprising several parts in a single module; that module can be replaced in its entirety, taken away and repaired in a factory environment and made available as a future replacement module itself. Often such modules will have within them simple diagnostic mechanisms which will indicate, perhaps remotely, that they have failed or are operating suboptimally, thus enabling the replacement module to be brought along ready for installation. If spares are to be kept they are likely to be stored at a cheaper warehouse location than at the premises where they are to be used, so there will be consequential wastes of transportation too.

There are some situations in which the provision of some spare capacity may be appropriate. For instance it is common for the main background level or baseline heating in a building to be provided by one boiler, with another cutting in to provide for peak loadings. There may be another boiler for when one or the other of those boilers is out of use, thus enabling the building(s) still to be heated to the normal level and for operations to continue in the buildings as normal. The provision of three boilers (and the space to accommodate them) could be considered a 'waste' (at least until it is needed!). Could building users be prevailed upon to work through undesirably cold conditions; should they? How would the costs of loss of production, if staff were to be sent home, compare with the cost of a spare boiler?

There are some areas where it may be difficult or not well advised to cut out inventory in the sense of spare provision. For instance, in a hospital it is important to be able to rely upon a constant power supply. Thus there will commonly be back-up generators and systems that cut in to ensure as far as possible an uninterrupted power supply (UPS). While hoping of course that the back-up system will never be called upon, it is an essential part of the provision, at least in the developed economies of the western world. In other parts of the world such spare provision may be considered a luxury when compared with other uses to which the capital funds may otherwise have been applied.

Motion

We have already considered the time and costs consumed in transporting maintenance operatives and materials from place to place; wasted motion occurs when people are required to expend more energy and time when, having arrived, they are actually doing the job. As so much building maintenance work involves small and not always repetitive jobs it is hard to build up any kind of time-and-motion data that would enable waste motion to be identified and eliminated. Simple things like having the right tools to hand rather than requiring repeat visits to the van, or the depot, to get them will go some way to address this. It may be, for instance, that it is worthwhile to get together assemblages or kits of appropriate tools and materials for various standard building maintenance occurrences.

There are some occasions when a 'while you're here' principle may need to be addressed. Does it make sense to do work additional to that anticipated and planned on the basis that it will save time in motions and transportation and waiting to do the work there and then rather than at a later date? It may, or it may not. Much will depend on what work if any is displaced or delayed as a consequence. Is the operative empowered to decide and to act accordingly; is higher authority required and is it available for consultation and approval at the time?

Overprocessing

This is about carrying out work to a higher standard than necessary. In building maintenance it is often difficult to determine the appropriate standard. Should the item after repair or other maintenance attention be expected to meet the standard that the item actually achieved (or was planned to achieve) when initially installed? Or the standard required of an equivalent new item today; or that deemed appropriate (in which case by whom?), bearing in mind the anticipated life of the building? Or for the item to still be expected to perform at a certain level so many years from now? For the item to perform at a higher level than that required is a waste. Such will occur if for instance three-coat paintwork is specified when two coats would meet the need – that over-specification involves overprocessing – more work is carried out than necessary.

Defects and rework

This is particularly expensive in relation to building maintenance. Egan (1998) identified that £1 billion was spent on rectifying defects on new buildings in the UK. That is a big waste. New buildings should be defect-free; why should the building client be expected to pay for something defective? Why should building operatives be paid for defective work? Atkinson (1998) showed that construction managers and operatives had the requisite skills and knowledge to build right first time but succumbed to pressures of time and money to cut corners and hope not to be found out. Construction managers, clerks of works and clients are also constrained by whether replacement workers could be found who would be more responsible and responsive to being checked up on. Building maintenance managers and operatives may find themselves initiating and carrying out work that ought not to have been accepted when the building was completed, and some of that will stem from inadequate design before that. It may be possible to seek to recover some or all of the costs of the rectification work; a decision will need to be taken of the likelihood of success.

Rework also occurs in relation to building maintenance work carried out. Sometimes this is due to poor or inadequate specification; often it occurs as a result of the one-off nature of much building maintenance. It also occurs because maintenance managers and operatives may feel constrained to make the minimum intervention necessary to keep the building functioning satisfactorily, only to find that the client or building user had higher expectations. Also, many building maintenance operatives are lowly skilled or unskilled; they may be styled as 'handyman'; others may be 'multi-skilled'. Such multi-skilling may be useful in that the operative may be able to resolve a range of inter-related issues, involving, for instance, fitting new cupboards, re-plumbing and redecorating. However the expression 'jack of all trades and master of none' may be apposite when a fully skilled person was really required for some part.

Other 'wastes'

A number of authors have sought to add to the Womack and Jones list of seven wastes. It is not necessary to expand in detail on these here, but to list a few possibilities only:

- underproduction, with consequential need to reorder and reset, missing out on a possible economy of scale;
- settling for a local but inferior product to reduce transportation;
- providing a prompt but inadequate response to reduce waiting;
- having inadequate inventory and thus losing production;
- having under-qualified or over-qualified staff;
- 'skimping' on doing the right job.

Standards have an inference of being a limit of acceptability. Those people, things, actions that meet or better the standard are acceptable; those that do not are substandard. British Standards are determined by committees of technical experts and practitioners in the area under consideration, advised and assisted by colleagues well versed and experienced in the construction of standards. Similar approaches are taken in the production of national standards around the world. In Europe there has long been a process of harmonisation, with common standards agreed as technical norms. There is a tension here between the setting of the highest attainable standard and the effect in lower-performing, poorer countries, where to introduce unrealistically high standards may result in higher costs or ignoring of the standards.

Customer service

Whatever may be deemed appropriate as national, or international, standards, it is arguably most important that the commissioners of building maintenance services should determine and set out for themselves just what it is that they want. It is then possible to monitor and measure the service being provided against that standard, which may or may not be formally incorporated in a contract document.

How might the required/desired/wanted/needed service be defined and designed, and the adequacy of its delivery be ascertained? Differences between 'want' and 'need' were discussed in Chapter 4; they are not discussed again here. However, whatever is defined in a contract must, as a minimum, be provided. It is important to be precise, to set out just what must be provided and to indicate clearly where there is scope for service to be provided with some flexibility. Some jobs may need to be carried out precisely in a particular way, or within particular times; others may have a lot of discretion attached to them. Generally speaking, the more discretion given, the greater the opportunity for the service provider to assemble works in an economical and efficient way and to bring forward new and alternative ways of addressing needs. On the other hand, the less prescriptive the customer is, the greater the scope for misunderstanding, surprise and dissatisfaction, with the potential for needing to undo and/or redo work. There will also be bad feeling, with dispute about who is at fault and who is to pay!

Prescriptive specifications set out precisely what is to be done and to what, defined, standard. They may be written or cut-and-pasted from standard specification clauses or from documents previously used, perhaps elsewhere, and hopefully updated and amended to suit the new situation. Such specifications may also prescribe who (or what qualification of person) should carry out the work, and when.

Performance specifications, by comparison, set out what it is that is to be achieved, without defining the way that it is to be attained. For instance it may be that a roof can be required that may be composed on any materials provided only that it achieves a U-value (a measure of thermal insulation performance) of no greater than 0.3 W/m^2K. While this approach has become quite common with newbuild, particularly with large building contractors and where there is scope for significant cost savings by 'shopping around' for alternative building products and materials, there is not so much scope in building maintenance work. By and large, maintenance work involves trying to match materials to existing, and often the methods to be used for repair reflect those of the original work. Hence there is often a need that particular work should be carried out by experienced operatives in whom the customer (and contractor) can have confidence.

It may be that customer expectations will be set out in a combination of prescriptive and performance specifications; they should be whatever is appropriate for the work to be done. There should be no (unpleasant) surprises. If the customer has set out clearly what he or she wants, then all the service provider has to do is to satisfy that. Is it that simple? Unfortunately not; for instance, customers are not always as clear as that. Nor are maintenance service providers all-knowing or suitably skilled or 'clean as the driven snow'. There is always scope for misunderstanding, so it is worth trying to recognise and minimise the scope for this and the effects that may ensue. For instance, in matters of health and safety, or working in occupied areas, it is probably worth being particularly precise and prescriptive about what is to be done, especially where the customer can be so. However, where the particular way something is done, or material to be used, is of little consequence, then there is less value in putting in a lot of expensive management time. But a lot of building maintenance work is uncertain at the outset as to just what is 'the problem', let alone what may be the (best) solution. In such circumstances a staged approach with preliminary opening up may be required to ascertain what is happening, in order to determine a more definitive prescriptive description of the required remedial work.

Do what the customer asks for and they will be satisfied; maybe. Many customers unfortunately only realise what they really wanted when confronted with what they now recognise is not that. (Real-ise = make real.) How that difference, between what was asked for and what was hoped for, is managed is critical to the relationship between customer and service provider, its quality and longevity. However tempting it is to refer to the contract that is often better left as a last resort – calling upon solicitors and going to court is time consuming, expensive and debilitating, and you may lose too! Goodwill takes time to build up but is quickly lost. So be sure to at least try to get it right, both first time and every time. When customer and service provider can trust each other, so much time, money and anxiety can be saved. When you make mistakes be sure to learn from them; better still learn from those of others.

So how to achieve customer satisfaction? At risk of sounding obvious, or to the cynical, radical, ask the customer. Bean (1997) and Smyth (1998) reviewed a system of client audit in which a large international construction contractor did just that. The contractor had been investing time and effort in trying to improve its performance in a number of areas it thought to be important to client satisfaction, only to find that clients' own priorities were other than those that the contractor had thought. Systems can be devised that are simple or complex, but they must start with asking the customer what they want. The customer will be feeling better disposed to the service provider from the time the contractor starts showing that degree of interest.

Some customers may not know what they want, they may be uncertain or may find it difficult to articulate their desires; they may feel intimidated, expecting that the contractor knows best (or should do) or that he may laugh. Examples of wishes, priorities, imperatives may be such as:

- Turn up on time.
- Don't turn up unannounced.
- Keep everything clean.
- Don't upset Chris.
- Stay within budget.
- Do what needs to be done.
- Don't keep stopping to ask.
- Don't press on without getting my OK first.
- Think: 'it's got to last 500 years' (an Oxford University college).

It may be useful for a contractor to have a small checklist or a number of prompts with which to initiate and develop a discussion. These may be based on the contractor's own priorities or matters on which they know they can deliver consistently well or on which they have been praised by other clients. It should then be possible to work towards a set of more refined and defined criteria or standards against which performance can be measured. Of course it is important to be reminded at this point that 'what gets measured gets done', that it to say that attention will be given to those issues above others. This can distort what happens in practice if not fully thought through. There was much attention given to this when targets were set in the National Health Service (NHS) in the UK, for instance in relation to waiting lists for operations; these were long and embarrassing politically in the 1990s and were to be reduced in size. Targets were set. This was most readily achieved by undertaking more of the short and uncomplicated procedures; lists reduced but waiting times went up for those who remained on the list – an unintended, unanticipated and undesirable consequential outcome. There can also be a tendency to count what can be measured and to ignore the more intangible.

Similarly, targets may be elevated to the position of Gods or idols, instilling fear and to be honoured at all costs. On 8th April 2008 Severn–Trent Water was fined a record £35.8 million for poor customer service and for giving the industry regulator (Ofwat) false information (BBC

News, 2008). The water company was allowed to charge its customers a higher rate if it met certain targets; it showed that it did. It could be said that there was incentive to cheat, to 'massage' the figures.

Some people relate readily to numbers. There is a simple assessment that can be made by comparing two or more figures – one is higher or lower than another and, therefore, better (or worse, depending upon what it is that is measured). Trends can be identified showing situations to be improving (or declining!). For structural and services engineers, quantity surveyors and accountants, numbers are their everyday currency; they are comfortable to make assessments based on 'the figures'. However, there can be thought to be limitations in such a reductionist approach, focusing on the figures rather than what is behind them. A poor result need not be seen as a failure or a sacking offence, but an indicator of an area to investigate further in order to examine and explain possible contributory factors and to address these positively.

If it is hoped to be able to report performance improvement and scale of improvement (e.g. substantial improvement) then data need to be collected and recorded, and to consistent standards. Such time series data can be used readily to show fluctuations and overall trends. They can also be used to inform possible targets for the future and the scale of improvement required, especially if performance is able to be benchmarked against performance achieved elsewhere in another organisation (or part) or at another time. A number of benchmarking clubs exist; some facilities management companies offer that service. It is common for like organisations, especially in the public sector, to routinely provide comparative performance data. For instance in the UK higher education sector the Association of University Directors of Estates (AUDE) collects and shares estate management statistics (EMS), for instance net and gross internal areas and functional suitability (AUDE, 2008). The NHS has a system known as ERIC – Estates Return Information Collection – which holds data on areas, function and space, quality of buildings, maintenance, energy, etc. While many measures of performance may be determined and defined by the bureaucrats or 'bean counters' for their own purposes, there is a developing area of interest in how the users of the facilities rate the service as they perceive it. NHS performance monitoring has seven main target areas at present – safety, cleanliness, food service, linen services, maintenance, environment and communications (NHS, 2008). There is an area on the website labelled 'Patient Power'.

Measures of satisfaction could relate to performance in relation to time, cost and quality, and include such factors as:

■ average time to respond to a maintenance request;
■ average time to return a defective item to required functionality;
■ proportion of maintenance requests satisfied within target time;
■ maintenance expenditure against budget, or compared with similar facility elsewhere;
■ courteousness/cleanliness/avoidance of disruption to satisfaction of occupant;
■ reuse of existing component in quest for increased sustainability (CSR agenda);
■ percentage reduction in complaints compared with previous year.

Some authors are speaking of 'customer delight' rather than satisfaction (Keiningham & Vavra, 2001; Wilson et al., 2001). Satisfaction suggests that the customer may be no more than 'satisfied' on a scale that may extend from very dissatisfied through to very satisfied, or from unacceptable through poor and satisfactory to fair, good, very good and excellent or outstanding. 'Satisfactory' may be no more than, say, a '2' on a six-point scale. 'Delight' suggests that a customer's expectations are exceeded and perhaps by some substantial margin.

There can be problems related to definition of such terms as 'satisfactory'; what is good enough for one person may be not so for another. The key here is that the customer decides for him or herself. So, arguably, even if the person who defined and commissioned the maintenance service is 'satisfied' with the standard of its provision, if users 'at the coal-face' in the workplace are dissatisfied, then the service is unsatisfactory. Terms of service should be reviewed for next time and performance improved to meet, and exceed, the higher threshold. Of course, service comes at a cost, so it may be necessary to renegotiate terms of payment and/or to discuss and agree what are reasonable expectations in the particular circumstances. It may be that 'Don't upset Chris' becomes a recognised (if not formally recorded) criterion.

A worked example of building a set of indicators of satisfaction or delight is provided at the end of this chapter.

Needs of the future: looking ahead

It is often said that expectations are rising; people are more demanding. Technology is increasing, in capability and complexity; 'Moore's law' says that computing power doubles every 18 months. Consider how cars have changed over the years. From manual ignition with a starting handle, farm wagon-like wheels with solid tyres and open, unheated saloons, we now have vehicles that almost drive themselves while their occupants travel in air-conditioned luxury. Our buildings too have changed, though some would say not so much; there is still much traditional construction. How might our buildings and their maintenance be different in the future? Is it possible or sensible to try to project and plan ahead. It has been said (Chapter 3): 'Fail to plan; plan to fail'. There is much wisdom in that, but it is difficult to foresee the future with any certainty.

A common approach is to plan the future by projecting from the past. Hence if over the last 10 years some aspect, e.g. sales or headcount, increased (or decreased) by an amount x, then it should be assumed that it will change again by x over the next 10 (or 5) years. That is trend planning. The communist regimes of the former Soviet Union (now Russia and a range of smaller states) and the People's Republic of China produced 'five-year plans' of agricultural and industrial production, planning out how many tractors, how much steel, would be needed in each of the years ahead. Public sector and commercial organisations in the western world too are still attracted to the idea for themselves, perhaps labelling it a strategic plan. Such plans tend to under- or over-estimate the scale, direction and nature of change. Is the future something one plans for or responds to? Who was able to foresee and plan for the decline in manufacturing industry and the growth of the former back office, the invention and take-up of air-conditioning, the desktop personal computer, the worldwide web and internet, homeworking, palmtops, Blackberries and social networking sites, such as *Facebook,MySpace* and *YouTube*? What next?

It could be your job, or part of it, to try to look forward and to plan. I suggest that if you see your task as one of responding to changing externalities that happen to or are inflicted upon your organisation, and somehow managing to accommodate to the situation, then that is what you will do; you will not have a seat at the boardroom table. If, however, you are seen as keen to help lead your organisation at the forefront of such changes, making change happen and driving through the change agenda, then you will be entitled to claim your key to the 'executive washroom'.

Can we project forward 50, 60 or 100 years? Some property decisions have long-term implications. For instance a property lease may commit the organisation to paying rent and other related charges for many years. That may be acceptable if the organisation and its products and markets are stable and they are able to stay there for the term of that lease and at a reasonable rent. However, if market conditions change radically the organisation may wish it had negotiated a shorter lease or a break clause so it could move to smaller or larger premises or to a building of higher or lower quality or in another location.

Can we project forward 5 or 10 or 20 years? What is the anticipated life of an office or shop fit-out? How much change can be expected; how much can be accommodated? Some buildings lend themselves better to reorganisation and alteration than others. Some organisations may be more open, amenable and accepting, even embracing of change. It is said that on the whole people are resistant of change. I suggest that the only situation in which you can be sure of no change coming is when you are dead. It is good, I believe, to try at least to accommodate to change even if one cannot in all conscience accept it as an improvement. Maya Angelou is credited with this quote: 'If you don't like something, change it; if you can't change it, change the way you think about it'; the Roman Claudiamus said 'Change or die'; and Mahatma Gandhi told that 'You must be the change that you want to see in the world' (www.worldofquotes.com).

Organisations may also be built, and their staff encouraged, to develop ways of working that are less resistant to and more receptive to change. 'Lean' and 'agile' are two words used to describe something of the underlying philosophy. If these are contrasted with 'robust' and 'solid', something of their attributes are suggested. Such systems are characterised by openness, delegation, originality and initiative rather than rules, regulations, paperwork and 'the way we do things here'. That is not to say that doing things well, taking time to consider actions and recording decisions are not appropriate; they are imperative whatever is done. An organisation which expects the unexpected rather than the regularity of routine is more likely not only to cope with change, but also to be proactive in promoting it.

At the time of writing, the business community is becoming increasingly sensitised to the need to show interest in, and action upon, the sustainability agenda. This interest in conserving resources, especially those in limited supply or non-renewable, has been growing since the 1970s. The then president of the Royal Institute of British Architects, Alex Gordon, wrote an article on 'resource conservation' (Gordon, 1974). Schumacher's book *Small is Beautiful* was originally published in 1973 (Schumacher, 1973), the World Commission on Environment and Development (WCED; Brundtland Commission) report, *Our Common Future*, was produced in 1989 (WCED, 1989) and there are now many books with sustainability in the title. With continuing interest in conservation, of heritage, of buildings and of energy, organisations must expect those carrying out maintenance to take and display interest in these matters, and to act accordingly. My own interest in sustainability gave rise to me publishing a paper on 'sustainable building maintenance' (Wood, 1999). I believe that interest in sustainability will continue, and increase, into the (so-called) foreseeable future.

What other issues or increases or innovations are likely to be significant drivers of change into that near future? By definition almost, the unforeseeable, that which may require and enable the greatest changes, cannot be foreseen or forecast. It can, however, be allowed for to an extent, though at a price. For instance the increasing height of the population could be allowed for by introducing taller doorways and greater floor-to-ceiling heights. Greater intensification of use of a building can be allowed for by having stronger floors, or shorter floor-spans, if so decided at the design stage. It would be expensive to strengthen floors once constructed and maybe prohibitively so. Moveable partitions are a way of resizing and reorganising offices in line with organisational changes. The most versatile provision for change is to have spare or excess space, but this will cost both in its initial construction or procurement and in keeping it in a state where it can be brought (back) fairly readily into use. On the whole, reduced demands can be accommodated, at least initially and for a while, by under-occupation, while future options of downsizing and relocating can be considered.

Risk assessments and sensitivity analyses can be carried out against alternative scenarios and projections of possible futures. The most extreme examples of change, while perhaps the most major in terms of impact and cost, may be amongst the least likely to come to pass. The effects of some changes will be countered or compensated for by other changes. For instance, an increased headcount may be accommodated by organisational changes allowing more home-working, together with reduced workstation size consequent upon replacement of desktop PCs with slimline screens on smaller desks. Whole product lines or services may be introduced or closed or relocated or outsourced and perhaps replaced by others as the organisation diversifies or returns to core business. If the organisation is able to produce a plan for its development and change over a period ahead then it may be possible to produce an estates strategy for the same period. Otherwise it must be expected that the estate will be adapted as and when proposals for organisational or operational change are presented and agreed.

> *. . . do not worry about tomorrow, for tomorrow will worry about itself. Each day has enough trouble of its own.* (Matthew ch.6, v.34)

A worked example – developing a checklist related to expectations

In recognition of the difficulty and importance of trying to determine let alone meet what may be expectations of a building or of its maintenance or of the maintenance team, I have included here a simple and straightforward two-step process.

Firstly you brainstorm what you think may be issues that may be of concern to the customer. These are just written on to a blank sheet of paper with no evaluation. They may come from anyone who may have an interest. They may be based in part on previous experiences, good and bad; there is no maximum or minimum number of items to record – some may be single words, others may encapsulate a consideration in a number of words.

The next step is to discuss with the client (ideally one person but it may be a small team) what these various words represent – are there some items which are almost the same thing or where one is a proxy or measure for another. The aim is to rationalise the items from the brainstorming into some kind of structured schedule with ways that it might be ascertained whether or not, or to what extent, the customers' needs or aspirations have been met or satisfied or not. The example shown here takes this step further in showing an agreed consequence of how much the customers' expectation has or has not been met. There is a column for 'LADs for lateness/dissatisfaction'. LADs (liquidated and ascertained damages) are a true measure of the loss to the customer resulting from the contractor's comparative non-performance. They are a predetermined amount that is a true estimation of loss, agreed between the parties and enforceable (should the need arise) through the courts. A penalty is not enforceable; it must be a reasonable and not punitive cost. It would be possible to extend this consideration to a possible bonus, if, for example a satisfaction rating of 6 were achieved on 95% of maintenance jobs.

Case study: maintenance delivery in a UK university

The university is located on three major campuses, one of which is in another county about 25 miles from the other two and has some buildings jointly owned with another institution; the university manages all the facilities.

Organisational context

Building maintenance is organised within the facilities management (FM) office which operates under a dedicated director; there are two deputies:

- A Head of Campus Services, who is responsible for:
  - ☐ three campus FM managers;
  - ☐ planned and long-term maintenance;
  - ☐ minor building projects;
  - ☐ sports and recreation;
  - ☐ transport;
  - ☐ safety and staff development.
- A Head of Estates, responsible for:
  - ☐ major projects and technical advice;
  - ☐ space management;
  - ☐ finance;
  - ☐ IT.

The Director of FM takes responsibility for catering and residential accommodation. The three senior managers together take responsibility for, among other things:

- estate strategy and objectives, service level agreements (SLAs) and key performance indicators (KPIs);
- property acquisition, management and disposal;
- capital works, procurement and project management;
- maintenance;
- business continuity and risk management.

As can be seen, several people have responsibilities for the determination and execution of building maintenance priorities and practices.

A campus facilities manager's (CFM's) role typically includes:

- supervising staff and/or contractors;
- recruitment, training, staff development and performance monitoring;
- ensuring optimum standards of customer care and cost effectiveness;
- organising and maintaining the helpdesk facility;
- preparing work specifications and briefing;
- safety and security services;
- budgets and record-keeping;
- capital works projects;
- statutory compliance and satisfactory standards of delivery;
- out-of-hours emergency maintenance;
- efficient and cost-effective use of campus facilities;
- regular reviews and benchmarking;
- promotion of good practice;
- effective liaison.

A 'typical' CFM is responsible for a complement of between 40 and 60 staff, including a maintenance team (either in-house or contract, of between five and seven) under a maintenance technician supervisor, a grounds team, a caretaking team and a security team, in addition to administrative support and domestic (cleaning and laundry) staff (Figure 5.1).

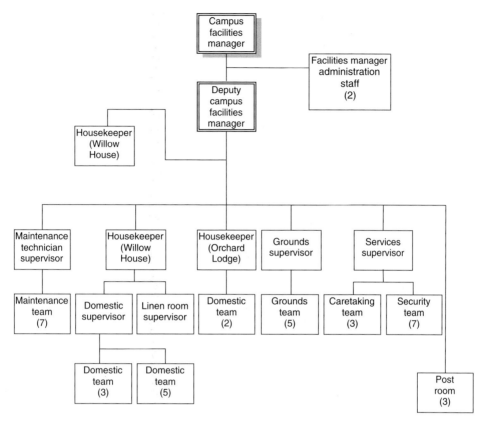

Figure 5.1 A university buildings and estates organisation chart.

Priorities

Contractors are required to carry out work in accordance with a University Code of Practice for Contractors; all personnel are to be appropriately trained and equipped, to wear uniform and carry an identity card at all times. A typical maintenance contract is of 3 years duration with an option to extend by up to two periods of 1 year subject to satisfactory performance in terms of quality and value for money. A '24/7' service is required, with defined response times:

▓ Priority 1: very urgent. Constitutes a danger or health hazard, or seriously affects occupation or operational effectiveness or security.
  □ *Investigate and make safe immediately; restore or provide alternative within 2 hours.*
▓ Priority 2: urgent. Failures or wants of repair that affect amenities but are not classified as very urgent.
  □ *Investigate, make safe and advise CFM; restore within 2 working days.*
▓ Priority 3: less urgent. Other failures or wants of repair that affect amenities but not classified as urgent.
  □ *Investigate, make safe and advise CFM; restore within 5 working days.*
▓ Priority 4: routine. Other failures or wants of repair not classified as priority 1, 2 or 3.
  □ *Investigate within one calendar week; repair within 10 working days after that.*

▪ The contractor's specified contact number shall be permanently manned.
   ☐ *Where a situation constitutes an emergency, the contractor shall attend within 1 hour.*

Repair requests are received by a central helpdesk and priorities are allocated by the maintenance technical supervisor for the relevant campus in accordance with the above criteria in the context of the workload as a whole; the supervisors have many years of technical experience gained working for the university. The key determinant of priority is the availability of back-up; for instance alternative toilet facilities close by. On some campuses work is done by in-house staff; on others it is outsourced.

Contractors are expected to become familiar with the requisite operating and maintenance manuals for each asset in order to facilitate satisfactory repair and operation of the installations in accordance with the contract. Prior to any work commencing, method statements and risk assessments and any subsequent modifications must be completed and approved by the CFM.

On completion, the contractor is to return work requisitions with a brief diagnosis of the defect, the work carried out and time taken on the task. These data are input to the computer system to inform management reports and future maintenance planning.

Monitoring and reporting

There is a 5-year rolling programme of planned maintenance, which is led by annual condition surveys of the buildings and mechanical and electrical (M&E) services. Planned maintenance schedules have been prepared for each building on two of the three campuses, indicating the activities to be carried out for each of the weeks of the year. Some tasks are scheduled to be undertaken weekly, others monthly and some quarterly, biannually or annually. For instance, emergency lighting and alarms are to be inspected weekly, whereas external lighting is to be checked monthly, refrigerant cooling systems and chiller controls quarterly, and external louvres disinfected annually. Monthly reports are required per property, to include:

▪ attendance of individual personnel on site, and when;
▪ recurrent items, for review;
▪ specific items of non-performance, and reasons therefor;
▪ forward projection of items nearing the end of their functional life.

Monthly liaison meetings are held between client and contractor representatives, with quarterly contract review meetings. Similar systems are applied where in-house staff are used.

The CFM monitors contractor performance against the '3 Es' – economy, efficiency and effectiveness. This includes assessments of continuing price competitiveness; compliance, responsiveness and reporting; quality of performance, service, advice and courtesy. Monitoring includes customer satisfaction surveys and complaints records, and performance is taken into account in decisions to extend or to terminate a contract.

Reflection

Operating with a range of response regimes provides a constant challenge. Historic organisational structures, perhaps not inappropriate in relation to the maintenance of historic buildings and the building up and retention of know-how, have been integrated as the university has expanded. However, this gives rise to different ways of doing things from one campus to another, and potential difficulties in learning from each other, transferring knowledge and building on best practice. Having said that, the university is operating effectively over a substantial estate and has good systems in place for planning and prioritising maintenance and for reviewing performance to inform future action.

Summary

This chapter has considered what may be reasonable expectations by the customer of a building maintenance service. It has examined various components of what may be considered as a satisfactory service and how that may be measured; it has also identified some limitations of measurement. Issues related to functionality and its loss and to the determination of appropriate standards have also been considered. Problems related to deciding what may be required of a building and its maintenance into an uncertain future have been debated and a view supported that even if the nature of actual changes cannot be foreseen, change can be accommodated. To assist practitioners to help customers determine and achieve satisfaction of appropriate expectations, a worked example has been provided.

The following chapter is designed to help customers, and practitioners advising and acting on their behalf, to set priorities across the multifarious and perhaps conflicting demands of building maintenance.

References

Association of University Directors of Estates (2008) *Estate Management Statistics.* www.opdems.ac.uk. Accessed 17 February 2009.

Atkinson, A.R. (1998) *The role of human error in the management of construction defects.* Proceedings of COBRA '98: RICS Construction and Building Research Conference, Oxford Brookes University. RICS, London, Vol. 1, pp. 1–11.

BBC (2008) *Record fine for Severn Trent.* news.bbc.co.uk/2/hi/business/7485006.stm. Accessed 17 February 2009.

Bean, M. (1997) *Developing and supporting a trail performance measurement system.* Proceedings of the 2nd National Construction Marketing Conference, Centre for Construction Marketing in association with CIMCIG, 3 July, Oxford Brookes University, Oxford.

Egan, Sir J. (1998) *Rethinking Construction.* Department of the Environment, Transport and the Regions (DETR), London.

Gordon, A. (1974) Architects and resource conservation. *RIBA Journal,* **81**(1), 9–12.

Keiningham, T. & Vavra, T. (2001) *The Customer Delight Principle: Exceeding Customers' Expectations for Bottom-Line Success.* McGraw-Hill, New York.

National Health Service (2008) *Estates Return Information Collection.* www.dh.gov.uk/en/Managingyourorganisation/Esatesandfacilitiesmanagement/index.htm. Accessed 17 February 2009.

Schumacher, E.F. (1973) *Small is Beautiful: A Study of Economics as if People Mattered.* Sphere Books, London.

Smyth, H.J. (1998) *The potential for realising innovative concepts for market segmentation derived from client audits.* RICS Construction and Building Research Conference COBRA '98, University of the West of England, Bristol.

Wilson, C., Leckman, J., Cappucino, K. & Pullen, W. (2001) Towards customer delight: Added value in public sector corporate real estate. *Journal of Corporate Real Estate,* **3**(3), 215–221.

Womack, J.P., Jones, D.T. & Roos, D. (1990) *The Machine that Changed the World*. Rawson Associates, New York.

Womack, J.P. & Jones, D.T. (1996) Beyond Toyota: how to root out waste and pursue perfection. *Harvard Business Review*, 74(5), 140–153.

Womack, J.P. & Jones, D.T. (2003) *Lean Thinking: Banish Waste and Create Wealth in your Corporation*. Simon & Schuster, London.

Womack, J.P. & Jones, D.T. (2005a) *Lean Solutions*. Free Press, New York.

Womack, J.P. & Jones, D.T. (2005b) Lean consumption. *Harvard Business Review*, **83**(3), 58–69.

Wood, B (1999) *Sustainable building maintenance.* Proceedings of the Australasian Universities Building Education Association 24th Annual Conference and catalyst '99, University of Western Sydney, 5–7 July, pp.129–140.

World Commission on Environment and Development (WCED) (1989)*Our Common Future*. Oxford University Press, Oxford.

Websites

http://dictionary.oed.com. Accessed 17 February 2009.

Worldofquotes.com. Accessed 17 February 2009.

6 Day-to-day prioritisation

The previous chapter considered expectations by the customer of a building and its maintenance. It examined components of what a satisfactory service and how that may be measured; it also explored some limitations of measurement. Issues related to functionality and standards were also considered. Problems related to uncertain futures were debated and a view supported that, even if the nature of actual changes cannot be foreseen, change can be accommodated. A worked example of how customer expectations may be identified, developed and documented was provided.

This chapter is intended to help customers, and practitioners advising and acting on their behalf, to set priorities across the multifarious and perhaps conflicting demands of building maintenance. It recognises that there is unlikely to be enough money or time or personnel to do everything that it is desirable be done, and that therefore priorities need somehow to be determined. It is suggested that sometimes money may be saved by putting off work, and that sometimes such deferral and delay may result in detriment and decay with consequential additional costs. The needs of today and tomorrow have to be weighed and a balance struck. Processes, systems, routines and rules of thumb are considered and a decision tree developed to assist in the determining of priorities, or to use a word disliked by some – prioritisation.

Basics

- Maintenance work that *must* be done, must be done.
- Work that must be done by a certain time must be done by then.
- Work that must be undertaken by people with certain skills and/or qualifications must be carried out by such, and verified.
- Where there are manufacturer's or installer's recommendations available to follow, they should be followed, or indemnities may be invalidated and liabilities incurred.
- A lot of maintenance work must be done *in situ;* due allowance must be made for travelling, including delays.
- Work that can be put off to another day or another year can be; but it may make sense to do it now – someone has to undertake evaluations, assess alternatives and make decisions.
- Maintenance work is a cost and an investment; it must be shown to be worth it.

Getting by

As indicated in the introduction, there is never likely to be enough money, time or personnel to do all that it is desirable be done, and therefore priorities need somehow to be determined. It would be possible to become very anxious about trying to weigh the various demands, some of which will be voiced loudly and perhaps unpleasantly. It is not worth getting anxious or angry; both are injurious to health and unhelpful.

A gentle answer turns away wrath, but a harsh word stirs up anger. (Proverbs ch.15, v.1)

What is needed is a way, or an armoury of ways, of eliminating, or at least reducing, anxieties and of taking the heat out of situations. Perhaps the use of the word 'armoury' is unhelpful, being suggestive of a battle of some kind in which people may be damaged and need protection. The words we use are powerful indicators of our attitudes, so it is worth trying to take time to consider what we wish to say, when, where and how, to whom, and what is the desired outcome. It will help if we are genuinely interested in the needs and concerns of those we are there to serve. If we are not addressing these needs and concerns, so that the situation is at least improved by our involvement, than what is it that we are there for? Doing the best we can with the resources we have is the best we can do; we should have peace of mind in that place.

Examples of words and attitudes that may help: Yes; I understand; of course; when can I come and see you?; what would you like to happen?; leave it with me, I'll get back to you and let you know what's happening; when would be a good time?

Words and attitudes less likely to help include: No; I'm busy; I don't have a magic wand; we have no money; I've got no staff; get back to me later in the year; you must be joking.

Putting off

Many requests (or demands) will come for maintenance responses that are immediate. To the person seeking to have maintenance attention, their need *is* immediate. The response they get is a measure of how important they and their welfare are perceived to be. They are not however necessarily party to the conflicting demands of various individuals and parts of the organisation. Nor are they likely to have the technical knowledge to be able to evaluate just what is the maintenance work that should be carried out. Nobody likes to be ignored or disrespected so it is important to be prompt and professional in response.

It helps if the organisation has an agreed statement or policy that will set maintenance decisions in a context. This will enable both the person responsible for deciding the priority of the particular maintenance job and the person initiating the request to understand the rationale behind the resulting decision, and why it may be necessary to wait for the desired action. From a financial point of view it may be thought to make sense to delay work until the latest time at which it can be carried out without incurring additional cost. For instance, delay will be counter-productive if a window that could have been repaired deteriorates during the period of delay to the point that it needs to be replaced instead. People located near the window may suffer draughts or be unable to open the window too in the intervening period; they will also have a constant reminder before them of how they may believe they are valued lowly by the organisation.

The more specialised the maintenance work the longer it is that the maintenance work may need to wait; there are fewer people qualified to do the work and they are likely to be committed already to a forward programme of work. Some specialist works, for instance in relation to gas and electrical installations, can only be carried out by appropriately qualified and certified individuals and organisations. The new job will need to be fitted into their programme at an appropriate time. The construction industry has been trying for some time to promote the idea of multi-skilling, such that an operative could perform a range of maintenance tasks. However, although it could be said that some of the trade and craft demarcations were close to restrictive practices, it has been difficult to break down the notion that such an operative may be 'jack of all trades and master of none'. The greater the multi-skilled availability, the quicker the response can be. A person could be fairly readily reassigned to a job with greater urgency; the deployment of the specialist is more constrained. A maintenance organisation will need to take account of such issues. The building in of slack or spare capacity can be very valuable in terms of speed of response, but is likely to be seriously questioned, and possibly curtailed, by the accountants. This has been a driver of outsourcing, where an outside organisation is contracted to provide the service. Certainly the larger the organisation the greater its ability to respond quickly and to provide cover outside normal working hours, such as evenings, weekends and holidays. There should then be less need to put off work, although it may be thought that the outside firm may not have the same interest in quality of service. On the contrary, that organisation is dependent upon giving good service if it is to have its contract renewed when the time comes. Its objectives will be to meet the same standards of service; they too will need to prioritise appropriately.

Maintenance and building inheritance

A significant amount of building maintenance can build up over time, creating a maintenance backlog. Personally I have some reservations about the term and the concept in that the organisation is still functioning satisfactorily; if it were not, and maintenance work not done were a significant contributory factor, then the issue would already be being addressed. I am happier to consider the so-called backlog a measure of difference between desires on the part of those requesting the work and those commissioning or sanctioning it. It can also be seen as a forward programme.

Existing buildings bequeath to those working in them and those responsible for their maintenance an 'inheritance' of work to be done. Some work may be presented as a backlog; other work may come forward as a feature of the kind of building that it is, its structural form and materials. For instance some reinforced concrete buildings may have problems of concrete cancer. Some buildings of the 1950s may have asbestos products within them, in roofs and cladding panels, in rainwater goods and as lagging. Some of these may be well documented, others may be 'lurking'. One of the best things a maintenance manager can do to help him or herself in doing their job well is to request or undertake a condition survey of the whole of the stock of buildings for which they are responsible. This will have several advantages:

- It will show that you are serious about the job.
- It will demonstrate to the board that such a survey is expensive and important.
- It will show that maintenance work to be carried out on the stock, expensive as it may appear, is but a small fraction, a single-digit percentage of the asset value.
- It will remind the board that the properties are an asset and need to be appreciated, cared for and working well for the organisation.
- It will enable you to prepare a forward programme based on how things really are and to justify your future actions and priorities.
- It will enable you to respond professionally to some who demand immediate attention when there are others in greater need (I nearly said 'give you ammunition' – hmmh!).
- You will be able to 'blame' your problems on your predecessors (for a while).

Heritage

Heritage is slightly different, though significantly, from inheritance (discussed above). Heritage is about *old* buildings, not just existing buildings, which may be comparatively new. Heritage conjures thoughts of ancient monuments, history, buildings of architectural interest and merit, 'monumental piles', historic town centres, conservation areas and listed buildings, architectural details and limitations on how they may be altered and kept up to date or just maintained or restored to some former glory.

In Britain, the Town and Country Planning Act of 1947 introduced the concept of listing buildings of special architectural, historical or cultural significance. Listed buildings need planning permission for their demolition and receive special attention in consideration of proposals for their alteration, normally requiring that original features be retained. The Civic Amenities Act of 1967 introduced similar protection for areas of cities, towns and villages where it was thought the character of the place should be maintained, through the creation of conservation areas.

In Australia, special protection is accorded to heritage buildings. Visiting that country demonstrated to me why it is important to retain that heritage, even though with my British perspective they seemed to me to be of hardly any age at all. There can be a tendency among those responsible for the maintenance of old buildings, and those living or working in them, to see their oldness as an inconvenience and a cause of greater cost. This is especially the case when coupled with the increased scrutiny and 'interference' from the authorities that comes with listed building or conservation area status.

As was suggested in Chapter 5, it may be as well to adopt that principle of Maya Angelou's: 'if you can't change something, change the way you think about it'. Heritage can be seen as a positive inheritance rather than inhibition. Often heritage buildings have a wealth of fascinating detail and history; they can be a joy, and a challenge, to investigate and to care for. If one can feel for the building as 'my baby' perhaps that state of bliss or thrill can be achieved, giving a level of achievement and satisfaction otherwise hard to attain. Such a building and responsibility may indeed be one to enjoy going to work for.

The helpdesk

I have not tracked down the first use of this term; it does not (yet, as at April 2008) feature in the Oxford English Dictionary; the entry in Wikipedia (wikipedia.org) describes it in terms of computing (perhaps a reflection of contributors to Wikipedia). It describes a helpdesk as 'an information and assistance resource that troubleshoots problems with computers and similar products'. The entry makes reference to an article by Middleton (1996). He identifies that the value of the helpdesk derives not only from response to users' issues but also the daily communication and related ability to monitor problems, preferences and satisfaction, valuable for planning. My tentative definition would be that a helpdesk is a virtual and/or real place or individual or group of people that receives queries and either answers them or responds to them, having sought advice or instruction, normally through a decision-support mechanism or group of advisors.

The principal beneficial feature of the helpdesk is that it gives a single point of contact. So all requests for building maintenance go to a single telephone number and/or email address. They are processed there in a way that accords priority according to a predetermined algorithm (set of rules) and the person requesting the maintenance action is advised of the response. They can be informed what will be done and when; and what the caller may be advised to do (if anything) in the meantime, perhaps to alleviate the situation.

The helpdesk puts a human and friendly voice and/or face into the otherwise perhaps fraught situation, taking out heat and anxiety and substituting salve and confidence that someone is 'on the case' for you. This is such an improvement on the norm of the 1950s when, if my parents wanted work done on their council house, they had to fill in a card, post it to the Housing Department, and wait for someone to call round, usually without appointment. There was no concept of 'customer care' let alone 'tenant power'. My parents had no idea where they were in the overall order of priorities or how soon, if at all, their request would be attended to.

Situations or systems in which priorities are determined by who shouts loudest or most often are unsatisfactory other than for the 'winner' and for the maintenance manager seeking the quiet life. In actuality, it is unsatisfactory for both of them too, because the 'megaphoner' becomes an unpleasant and angry character, and by giving in the manager invites all in need to resort to labelling all their requests 'urgent'. The professional maintenance manager will see it as an important part of his or her role to determine and distinguish between urgency and importance, and to order work accordingly, balancing want and need, supply with demand, and today and tomorrow.

Today's needs

Arguably the most important aspect of building maintenance is to enable the continuing and efficient operation of the building for whatever purpose the building is held. Thus if the building is for the production of widgets then it must be maintained in such a way as to enable that to be done. If the building is to enable the production of widgets Monday to Friday 9–5 that is an easier task than if it is to do so 24/7. If the building is for the handling of the administration of the organisation that produces widgets, then its maintenance is a different task again. If the building contains production and administration, then its maintenance is a more complex task, though it may be easier to carry out. The key task is to be able to continue operations at the time(s) when that is expected. Loss of production means a serious reduction in profits. If a firm is expecting to make a profit of say 5% and it loses even 1 day of production while still incurring costs, that 0.5% of production loss will impact directly on the 'bottom line' with a 10% reduction in profit to now only 4.5%.

It is said that 'time is money'; time lost is lost for ever – it cannot be created or recreated. There is sense in which there is only today. Yesterday has gone; there is nothing can be done with that other than learn from it; there is limited time in which to do that and it may be more profitable to 'move on'. Tomorrow however, while always beckoning, never comes.

Tomorrow

What will the future be? Who can foretell? What will be the building maintenance context?

The World Bank (2007, cited in Moynagh & Worsley, 2008) projects that in 'rich countries' the average income will rise over 20 years from about £15 000 to about £25 000 at 2007 prices. This doubling in purchasing power is more or less the same as has happened in the past 25 years. At the same time there is a mass of cheap labour worldwide; expansion of the European Union, increased mobility and globalisation has changed the geo-economics of labour. Who can tell what that will mean for buildings and their maintenance into the future? Emerging economies tend to grow faster than established ones; there are new growth centres. Companies may relocate; buildings will have changing values and uses. Such changes will impact upon building maintenance.

Possible future scenarios

Trying to forecast the future is fraught with problems; trying to plan for it, even more so.

- ■ What skills and knowledge will be available; what will be required?
- ■ Where should our buildings be located? What proximities will be important, if any?
- ■ How important will interpersonal relationships be: teamwork; transparency and trust; customisation/personalisation?
- ■ Will there be greater virtuality, offshoring, migration, mobility – or less?
- ■ What will be the effects of global warming and climate change; availability and cost of energy and water?
- ■ What will be the economic climate? Greater stability or less for planning ahead? More regulation, or less? More 'organisation' – accountability, best practice; monitoring, evaluation, audit; reorganisation?
- ■ Will there be more work, or less; more leisure? More workers or fewer? More managers or fewer; 'delayering'?; more training/retraining? Fads?
- ■ How will technology, automation, ICT, solar heating, wind and tidal power, help?
- ■ When will the future come? How will we know? What is already happening elsewhere?
- ■ What will be the effect if we plan for the 'wrong' future, a future that doesn't come? What if our plans are inadequate; if the future is worse than we assumed?
- ■ We could spend a lot of time and money contemplating and planning for the future, including worst case scenarios that probably won't come to pass; but they might!

In 1992 about 20% of the top 50 UK companies were engaged in large-scale reorganisation each year. This had jumped to 30% by the end of the decade. (Wittington & Mayer, 2002; cited in Moynagh & Worsley, 2008, p. 141)

More than nine in 10 of 2000 workplaces surveyed in 2002 had outsourced at least one type of activity, and one half had outsourced four or more. (Moynagh & Worsley, 2008, p. 142)

These are considerations that may be had at a macro level; they will also impact what happens at lower and local levels. The future will have its way.

The work required tomorrow might be different from that required today. It may be, for instance, that an element that was to be repaired has incurred further damage or decay, or that legislative requirements have been amended, or that the needs of the business or organisation have changed. It may also be, of course, that the work understood to be required was not quite that which would be most appropriate or beneficial in any case, given further consideration. Matters related to deterioration and defects, their identification and rectification will be dealt with in subsequent chapters.

It is conventional to divide works into major and minor works, capital projects and revenue expenditures, day-to-day response or longer-term planned maintenance programmes. The divisions can be hard to define and to make, but we must try.

Major works

Almost any new building other than a store or shed could be considered a major work. Similarly, large refurbishment projects could be considered as major works. In essence anything where a building or a substantial part of a building is transformed or renewed, such that it might not require further major work for many years ahead. Major works are of such a size that will almost certainly be funded as capital projects (see below) and planned. Sometimes, major works are defined by their size or value. For instance, in Australia, the government of Tasmania defines major projects as 'capital investment projects, including construction and maintenance, which have an estimated total value greater than $100 000' (Tasmania Department of Treasury & Finance, 2007).

For major works there will almost certainly be a contract between the building owner and a building contractor. The choice of contract is discussed under Procurement below.

Minor works

By contrast, minor works are small: they may range from replacing a missing screw (though it would be good to also wonder why the screw is missing), or rehanging a door or gate, or several such small works, up to the small building. There is a large range of what may be construed as minor works. Often organisations will have agreed limits of what may be so considered; I give three examples here from different sectors.

The National Health Service (NHS) has, over recent years moved from being a huge, bureaucratic, single organisation to a service delivered through a multitude of locally based NHS trusts. Whilst giving greater autonomy and ownership at the local level this has resulted in some loss of benefit of scale. The central organisation is more focused on strategic planning and advice. One such initiative is the ProCure21 programme, aimed at achieving better procurement of building works. Within this a minor works contract has been devised, 'for schemes under £1 million'.

> Traditionally, minor works can be costly and time consuming to administer. Each task has to be managed on an individual basis and presents a serious drain on NHS resources. The ProCure 21 Minor Works contract allows the NHS Trust to manage all tasks under one contract with one contractor. (NHS, 2008)

In other institutions, financial limits may be much lower. For instance, Leeds University expresses it this way:

> The Minor Works Fund is an annual allocation of £500,000 ... It is available to support departments/schools who have need for minor works up to a cost limit of £50,000. (Leeds University, 2008)

The Church of England, an institution of even longer standing, and having a substantial stock of very old buildings, many of which are of great historical value, has more restrictive policies regarding minor works (Diocese of London, 2008):

> The following minor works do not require a faculty[1] although it is recommended that churches contact their Archdeacon or the Care of Churches Team before starting work ...

> 1. Works of routine maintenance on the fabric of the church (not materially altering its appearance) including scaffolding up to £5000 (excluding VAT, and excluding the cost of any scaffolding) ...
> 2. Works of routine maintenance of electrical fittings or other electrical equipment (by approved NICEIC electricians) and furniture up to a cost of £2000 excluding VAT ...

In many ways the execution of minor works can be as involved, if not more so, as major works, and almost certainly disproportionately so. The Joint Contracts Tribunal (JCT) has a form of

---

[1] Faculty: a formal application and approval system that may take several weeks or months to complete, and may require specifications, drawings and tenders to be completed.

contract specifically for minor works, and this is discussed under Procurement below and in Chapter 9.

> *The Minor Works form of contract is appropriate where:*
>
> ■ *the work involved is simple and clear;*
> ■ *drawings and/or specification and/or schedules to define adequately the quantity and quality of the work; and*
> ■ *an Architect/Contract Administrator is to administer the contract.*
>
> *It is not suitable where Bills of Quantities are required or where detailed control mechanisms are needed.* (JCT, 2005)

The Intermediate form

Recognising the difficulty of determining and differentiating between major and minor works, and the range of ways that such works may be best be organised, JCT has prepared an Intermediate form of contract; this is also discussed below and in Chapter 9.

> *The Intermediate Building Contract is appropriate where:*
>
> ■ *the proposed building works are of simple content involving the normal, recognised basic trades and skills of the industry, without building service installations of a complex nature or other complex specialist work;*
> ■ *fairly detailed contract provisions are necessary and the Employer is to provide the Contractor with drawings and Bills of Quantities, a specification or work schedules to define adequately the quantity and quality of the work; and*
> ■ *an Architect/ Contract Administrator and Quantity Surveyor are to administer the conditions.(JCT, 2005).*

Capital projects

Capital/revenue is the other classical divide of construction-related work, though it can be difficult to draw a line. HM Revenue and Customs (2008) states:

> *What is and what is not capital expenditure has taxed the minds of judges, tax advisors, Inspectors, business people and others for more than two centuries. No one has produced a single, simple test that will determine the issue in all circumstances . . . The day-to-day running costs of a business (staff wages, purchases of trading stock, rent of business premises, and so on) are referred to as revenue expenditure.*

Here is a definition of 'capital project':

> *A long-term investment project requiring relatively large sums to acquire, develop, improve and/or maintain a capital asset (such as land, buildings, dykes, roads).* (Business Dictionary, 2008)

Here is a definition of a 'capital project' from a local authority in the USA:

Sometimes budgets will be cut part way through the year if the organisation's income is not as planned. This can create difficulties when some monies have been committed to maintenance projects, leaving the cuts to fall entirely on funds uncommitted, which will typically include emergency and breakdown maintenance. It can therefore be useful to have some contracts which are spread over more than one financial year and to draw down monies over the year or years against a monthly plan. This also enables plans to be adapted in the light of changing circumstances, whether good or adverse.

Costs must be contained. Jobs invariably cost more if there are many unknowns about them. It is cost effective to have long-term contracts that spell out as clearly as possible the work to be done, how and when it is to be done, when that is important, and to leave the contractor as much scope as possible to work efficiently in his or her own way. To allow a contractor to work on a 'daywork' basis, submitting timesheets for work of uncertain specification and taking as long as it takes is to open the client to unlimited expenditure. Ideally, all work should be done to agreed specifications and for agreed sums of money. A contract sets out such terms.

Contracts

A number of forms of contract have already been referred to in this chapter:

- the Minor Works form;
- the Intermediate form;
- the Standard form of building contract.

These three are issued by the JCT, a corporate body representing building contractors, professional bodies and latterly client bodies. They are intended to offer fair distribution of responsibilities and to attribute risks where they are best able to be taken. The JCT website gives details of the contracts' attributes.

Other forms of contract are available. For instance the Chartered Institute of Building offers a suite of contracts including:

- Minor Works;
- Small Works;
- Mini Form of Contract;
- Facilities Management Contract.

NEC, set up to promote the New Engineering Form of Contract, offers the Engineering and Construction Contract (ECC) Term Service Contract. There are various options, e.g. cost reimbursable, framework contract, target contract; and there are guidance notes available.

The Association of Consultant Architects promotes PPC 2000, The ACA Standard Form of Contract for Project Partnering.

The JCT Measured Term Contract is especially apposite. It is appropriate for use by employers with a regular flow of maintenance and minor works to be carried out by a single contractor over a specified period of time with work instructed from time to time. The work is measured and valued in accordance with an agreed schedule of rates (e.g. per square metre, or each) with a contract administrator and quantity surveyor engaged to administer the contract. A common schedule is the PSA Schedule of Rates for Building Works 2005 published by The Stationery Office. This has a provenance of long-standing, having been initiated by the UK government's then Property Services Agency and continued by a major provider of facilities management and maintenance services, Carillion. The PSA schedule lists over 20 000 rates and is regarded as the standard for the public sector (Carillion, 2005). The Society of Chief Quantity Surveyors (SCQS) has issued a code of practice related to building maintenance using national Schedules of Rates (SCQS, 2007).

Procurement

How to go about obtaining the maintenance service required? Firstly, and at risk of re-stating the perhaps obvious, be sure you are clear about what service *is* required.

Now you need to ask, and answer, a number of questions:

Are there new-build projects to be undertaken? Are they similar to works carried out before; satisfactorily? Do you see a forward programme of similar work? How specialised are they? Should (could) they be carried out concurrently with other works, of the same nature/in the vicinity? How large are they; how long are they likely to take? How prescriptive is the client about how the work should be done and how long it may take? Is an architect or surveyor to be involved? Is planning permission/listed building consent required; has it been sought; obtained? If similar work has been undertaken then it may be sensible to negotiate similar (and possibly better) terms with the same contractor? Both parties will have less risk.

Are there large-scale refurbishments to be undertaken? Similar questions to those above can be asked. Will particular heritage skills be required; are they available? Is this work of such a nature as to suggest a long-term relationship may be beneficial? A particular issue with refurbishment work is the extent of additional work that can only be specified once areas are uncovered; they are often worse than anticipated. It is wise to allow appropriate contingency sums to cover such eventualities. It may be that all the major types of work can be specified and measured for a contract to be tendered for or negotiated and to allow a contract price to be determined. The work can then be re-measured on completion and the contract sum varied accordingly.

Is the work to be carried out a small number of similar repair/replacement jobs – say plumbing works? If so, a specialist firm can be engaged. If there is work of a variety of kinds, needing for instance plumbers, carpenters, roof tilers and electricians, it may be better to engage a general contractor on a one-off basis. If, however, it is possible to foresee a need for such over a period of time, perhaps 2 or 5 years, or more, it may be sensible to seek tenders against a Schedule of Rates for a measured term contract. A break clause could be inserted to allow for termination in the event of dissatisfaction on the part of either party; and/or an extension could be negotiated in the event that both parties so wish.

Some clients may see benefit in negotiating a long-term relationship with a contractor that could extend from design, financing and construction of the building to its fitting out, day-to-day operation and maintenance, and continual/periodic upgrading. Clearly such an arrangement will need very careful consideration and contracts will need be so drawn as to cover as far as possible all foreseeable eventualities and to have mechanisms in place to deal satisfactorily with the unforeseeable! Such contracts are only likely to be able to be entered into by large corporate organisations with solid expertise and experience; those without such backgrounds would be well advised to seek, and act upon, impartial professional advice from people who have. 'A fool and his {/her} {organisation's} money {/job} are soon parted.'

Decision making and recording

Whatever pathways are taken to deliver the repair or replacement decided upon, it is important to be methodical in the taking of that decision and in the making of records of such decisions.

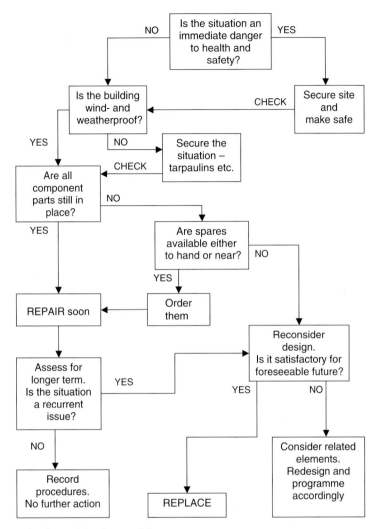

Figure 6.1 A simple repair/replace model.

Figure 6.1 shows a decision-making tree. In this example, the first level of consideration and decision is whether to repair or replace the component, element or building. Firm criteria against which to decide will help. If the decision is to replace, then a further consideration must be made and decision taken as to whether the element should be replaced on a like-for-like basis or with something different. The replacement should be monitored and its performance – whether, for instance, better than previous, or a disappointing outcome – should be recorded in a database and used to inform future decisions. Similarly considerations should be made for repair decisions, for instance to apply a quick fix or to make a more thorough repair. The process allows for effective decision-making processes to be developed and thereby good decisions to be made and put into effect.

Summary

This chapter has outlined a number of considerations that need to be made in deciding priorities across the multifarious and conflicting demands of building maintenance. It recognises that there is rarely enough money or time or personnel to do everything, and that therefore priorities need somehow to be determined. Whilst money might be saved by putting off work, such deferral and delay may result in detriment and decay with consequential additional costs. The needs of today and tomorrow have to be weighed and a balance struck.

The following chapter considers the deterioration and decay of buildings and services over time.

References

British Standards Institution (1998) BS 4778-3.1: 1991 (updated 1998) *Quality Vocabulary – Part 3: Availability, Reliability and Maintainability Terms – Section 3.1 Guide to concepts and related definitions.* BSI, London.

Business Dictionary (2008) www.businessdictionary.com/definition/capital-project.html. Accessed 18 February 2009.

Carillion (2005) *PSA Schedule of Rates for Building Works* (19th edn). The Stationery Office, London.

Cheshire County Council (2008) Capital and Revenue Expenditure: Definitions. www.cheshire.gov.uk/Council/Constitution/Part4/FinanceProcedureRules/Annex1.htm. Accessed 27 February 2009.

City of Portland, Oregon (2008) Definition – What is a Capital Project? http://www.portlandonline.com/omf/index.cfm?c=30079&a=50495. Accessed 18 February 2009.

Diocese of London (2008) www.london.anglican.org/DACMinorWorks. Accessed 18 February 2009.

Finch, E. (1988) *The use of multi-attribute utility theory in facilities management.* Proceedings of Conseil International du Batiment Working Commission W70, Whole Life Asset Management, Heriot Watt University, Edinburgh.

HM Revenue and Customs (2008) Business Income Manual BIM35005 – Capital/Revenue Divide: Introduction: What is capital expenditure: Historical review. www.hmrc.gov.uk/manuals/bimmanual/BIM35005.htm. Accessed 27 February 2009.

Joint Contracts Tribunal (2005) *JCT 2005 Suite of Contracts.* www.jctltd.co.uk. Accessed 18 February 2009.

Leeds University (2008) www.leeds.ac.uk/estate_services. Accessed 18 February 2009.

Middleton, I. (1996) *Key Factors in Help Desk Success (An analysis of areas critical to help desk development and functionality).* R & D Report 6247. The British Library, London.

Moynagh, M. & Worsley, R. (2008) *Going Global: Key Questions for the 21st Century.* A.& C. Black, London, p. 29.

NHS (2008) www.nhs-procure21.gov.uk/content/minor_works.asp. Accessed 18 February 2009.

Parkinson, C. Northcote (1958) *Parkinson's Law: The Pursuit of Progress.* John Murray, London.

PJB Capital Allowances and Depreciation Consultants (2008) *Datasheet 04-2.* www.pjb.com/pdf/datasheets/datasheet04-2.pdf. Accessed 18 February 2009.

Saaty, T.L. (1980) *The Analytic Hierarchy Process.* McGraw-Hill, New York.

Society of Chief Quantity Surveyors (2007) *Building Maintenance using National Schedules of Rates – A Code of Practice.* SCQS, Huddersfield, UK. www.scqs.org.uk. Accessed 18 February 2009.

Tasmania Department of Treasury & Finance (2007) *Conventions and Glossary of Terms.* http://www.treasury.tas.gov.au/domino/dtf/dtf.nsf/d4790d0513819443ca256f2500081fa2/ca7a076d89f716f9ca25718a001237f7?OpenDocument. Accessed 18 February 2009.

Rainwater may also cause problems in relation to foundations, specifically subsidence due to undermining of footings. This may relate to rainwater that has fallen on the building and has been concentrated by the rainwater disposal system into a place where it can cause a problem. For instance, a blocked and cracked downpipe or gully may cause the rainwater to discharge around the outlet and to wash away soil from under a barely adequate foundation. Problems may also arise from groundwater coming on to the property from higher ground or from surface water unable to clear quickly enough due to saturation of the ground. Anecdotally at least, rainfall intensities seem to be increasing, giving rise to stress on rainwater disposal systems designed on the basis of lower loadings and therefore with capacity inadequate for the current need. This should be reviewed as circumstances may suggest.

Today, consideration should also be given to rainwater 'harvesting' and mitigation. For instance a sedum roof (sedum being a low and slow-growing succulent plant) may take up water and slow (attenuate) its progress, thus helping reduce risks of overloading the drains and consequential flash flooding. The roof structure will need to be assessed for its strength and surfaces checked for their completeness. Also, rainwater storage tanks could be installed for the same reason and to allow for gardens to be watered therefrom. Exceptionally, rainwater could be collected and used to flush toilets or even purified for use as drinking water. In this way, rainwater falling on a building could be seen as a positive virtue, and valued, rather than a problem to be dealt with and disposed of.

Damp

Dampness may be related to precipitation, but it might not be. Processes may be similar but different. For example, some of the situations described above in which water may enter the building, such as at dormer windows, may be manifested by dampness internally. More detailed investigation and analysis will need to be carried out to ascertain the actual cause or contributory causes. These situations will be considered in more detail in subsequent chapters. Two possibilities other than penetrating damp will be considered here: condensation and rising damp.

Dampness will be made manifest by such as:

- peeling wallpaper;
- cracked paintwork;
- cracked, hollow or crazed plaster;
- rusting metal;
- rotting timber;
- mould (grey, green, blue, black, white).

(Yellow, orange or red powdery deposits or stains are more likely to be related to rots and decay in timber – these will be discussed in Chapter 14.)

Condensation

Condensation in a building occurs in the same way as rain is formed in the air. When air is warmed it is able to hold more moisture, and when it is cooled it can hold less. When the air cools below its *dew point* the moisture within it precipitates out as condensation. In buildings this condensation is most commonly manifested as moisture on the inner face of the glass pane of a window. Single glazing is more prone than double- or triple-glazing as the glass surface is cooler; and kitchens and bathrooms are more prone as those rooms have more moisture generated in them than other rooms. That is not to say that other rooms and other surfaces are not also vulnerable – they are! The key is the combination of a cool surface and excess moisture. In the same way as winds are generated externally, warm moist air travels to displace colder air and thereby reaches the associated cold surfaces. If the surface is impervious, or relatively so, such as gloss paint, the condensation will be readily visible on the surface. Other kinds of decorative finish, for instance wallpaper (and especially woodchip paper, often used in domestic refurbishment projects) may become dampened by the condensate, and the combination of mould spores and moisture can result in very unattractive mould growth. This will generally be seen first at window reveals and in room corners, which are the coldest parts of rooms and have least air movement, especially where furniture is inhibiting circulation. In bad cases the mould can also affect furniture and contents. Such mould, whether or not it is considered unhealthy in itself, is certainly a manifestation of an unsatisfactorily damp internal environment.

Dampness caused by condensation may also occur unseen. Where the temperature gradient across an external wall (typically warm internally and cold, perhaps below freezing, externally) is such that the dew point occurs *within* the wall structure, then moisture may condense out interstitially, that is to say within the interstices of the wall. This can cause unseen deterioration.

In general terms, the answer to condensation may be found in improved heating and/or insulation and/or ventilation. The Building Research Establishment (BRE) produced an excellent guide, *Thermal Insulation: Avoiding Risks* (BRE, 1989). Note that the title does not suggest that *all* risks will be avoided, only that some may be – indeed 'reducing risks' may be more accurate. Why was that publication produced at that time? It corresponded with an 'epidemic' of mould-related manifestations of condensation in large-scale housing, both newbuild, in the form of mass industrialised system building, and in large-scale refurbishment of older council estates. Many of the industrialised systems featured large panel systems (LPS) with concrete walls and roofs with a thin layer of thermal insulating material sandwiched within the precast concrete elements. Generally the overall thermal insulation was poor (often the insulating material was incomplete or inadequate when submitted to testing) and prone to both surface and interstitial condensation. In the refurbishment projects, often well intentioned improvements to thermal insulation of external walls resulted in condensation and mould problems as those walls became colder than previously. In other instances, where, for instance, new extensions were built to modern constructional standards, condensation problems were created in the existing, less well insulated structures – a case of solve one problem and create another.

Often insufficient account is taken of how buildings may or will be used. For example a housing refurbishment may be carried out with thermal insulation improved and a central heating system designed to provide room temperatures of 20°C when the outside temperature is −1°C. All may not be well if the occupant is unwilling or unable to afford to run the heating due to high fuel costs. This is a circumstance known as fuel poverty, defined in UK as a situation where the occupants are spending more than 10% of income on energy. Condensation and mould may be expected when the building is not used as designed. The constructional form also needs to match the use of the building. For instance a house of traditional masonry construction functions well if it is fairly continuously occupied, such that heating inputs during the day are stored in the structure and keep the house warm through the cold night. However, a building that is occupied intermittently, and only during the daytime, may benefit from a lightweight structure that will warm up quickly when heating is switched on, and similarly cool down quickly, so that expensively produced heat is not wasted. Social changes in UK during the 1960s and 1970s resulted in fewer families at home during the day and more people at work in offices. Building designs and constructions needed to accommodate to the change, and sometimes they didn't. As fuel prices are again increasing rapidly we may find similar processes at work.

The causes of condensation are complex and consequences can be unpleasant, so it is well worth taking time to consider the problem.

Rising damp

Problems of water ingress from outside through ineffective roofs and gutters and so on may be considered as falling or penetrating damp. Ceilings may be affected by penetrating dampness or surface condensation. Walls may be similarly affected; they may also be affected by rising damp. This occurs when the wall is in contact with moisture in the ground. The commonly accepted theory is that water in the ground is drawn up into and through the wall structure by capillary action in pores in the wall materials. Dampness will often be found at the bottom of walls in contact with the ground. Over the years since the Victorian period the damp proof course (DPC) has been developed. Initially this was a layer (or two or three) of slate or engineering brick, intended to resist the upward passage of moisture from the ground. DPCs and related defects are discussed in more detail in Chapters 11 and 12; suffice it here to say that the key quality required is of *completeness* of the DPC; it should not be broken or bridged. Bridging occurs where material such as render or plaster crosses the DPC. Rising damp will often be seen on the lower metre or so of the ground floor of affected buildings, where the dampness will affect plaster and decorations in the same way as other forms of damp.

Wind

Wind is created by air moving from an area of high pressure toward one of low pressure. The speed of air movement is a factor of the amount of difference in pressures. Air pressures are shown on weather maps by isobars – lines linking places of the same barometric pressure. The closer together the isobars, the more intense the wind.

Wind speed is also affected by terrain. In rural areas, buildings may be sheltered from winds by hills, trees and hedges. Similarly winds may be funnelled, and thereby intensified, by valleys. In urban areas, buildings may have similar effects on each other. At low levels, winds are generally slower, being disrupted by many buildings, although there may be 'canyon' effects generated in some areas. At higher levels, there are fewer buildings to slow the wind, but flows can be seriously disturbed. Typically, wind speeds are increased at corners of buildings, so these are more vulnerable, both to air infiltration and to associated water ingress under the pressure of driving rain.

Wind can also have effects on the *leeward* (sheltered) side(s) of buildings, where suction is induced. Thus not only does the external envelope of the building need to resist the pressure of wind blowing on it, it also needs to resist pull-out forces that may also be substantial.

In most locations there is a direction from which the wind most commonly comes; for instance in the UK the prevailing winds come most often from the southwest. That is however a generalisation. For example, on the east coast, easterly winds arriving from a cold continental Europe over a cold North Sea can be both very cold and strong.

Winds do not blow at a constant speed; they may gust, that is to say to blow for a time at a much greater speed. These gusts can be particularly destructive, for instance resulting in the stripping of roof tiles or even entire roof structures.

While this book is not intended to give a substantial treatise on weather, it is hoped it will sensitise the reader to weather-related issues that have substantial importance for the deterioration and maintenance of buildings. Walls of tall buildings and roofs generally are particularly prone to wind-related problems and these are discussed further in Chapters 13 and 14. The BRE has produced much excellent advice on designing buildings to take account of wind. Data is available for locations across the UK showing wind 'roses' of how much wind arrives from the various points of the compass and their intensities; maximum recorded wind speeds are also available. In some parts of the world, wind-related phenomena such as tornadoes need to be taken into account, with their particularly intensive winds and destructive potential. Wind can be a major factor in deterioration.

Sun/ultraviolet light

The sun can also be a major factor in deterioration of parts of a building. The sun is generally at its most intense at noon and in the hours thereafter when it is at its highest, the air has warmed and atmospheric pollution dissipated; in the northern hemisphere this will occur when the sun is in the south – and conversely in the southern hemisphere. The sun can heat surfaces and spaces within, especially roofs and roof spaces, intensely, for instance causing bituminous materials to expand, melt, move, blister and crack. Metals will also expand and contract as temperatures rise and fall.

The sky need not be cloudless for the sun to have its deleterious effects. Much of the damage is caused by ultraviolet (UV) rays, which are radiated irrespective of cloud cover. UV can damage paintwork, dulling its gloss and causing cracking and splitting of both paint and the material beneath. Plastics become brittle under the influence of UV.

The sun can also have serious effects on building occupants, affecting their working conditions and thereby productivity and satisfaction. There can often be substantial overheating in areas on the 'sunny' sides of buildings. Although not perhaps a 'maintenance' item to some it is an important matter for those responsible for achieving satisfactory internal environments. The solar environment may have been given insufficient attention at the design stage; if so, this will need to be dealt with during occupation. Mitigating measures may include such as external or internal shading devices such as blinds, *brise soleil*, and materials to be applied as a film to window glass. Shading, however, may also bring problems – for instance shaded areas of material will expand less than those that remain exposed to the heat of the sun; differential movement may cause cracking of glass unable to accommodate the differential stresses within.

Temperature changes

Temperature has an effect on building components. Although commonly the effect may be small and maybe even negligible, there may be situations in which there is a noticeable effect. This may be more noticeable when temperature changes. For instance plastic guttering expands in the heat of the sun and contracts when the sun 'disappears' behind clouds – this can be accompanied by noticeable creaking sounds. The material is unlikely to be affected in any long-term way provided that the component parts have been fixed correctly so as to allow for the thermal movement.

Exceptionally, however, there may be cases where localised temperatures become very high due to particular configurations. For instance a plastic or rubberised edging or skirting may become permanently distorted by the effect of concentrated sunlight adjacent to glazed areas or heating pipes. Such heat may also dry out timber, causing it to shrink, opening up unsightly gaps in floors, skirtings and panelling. When a building is first occupied, or reoccupied after a period of disuse and/or refurbishment, the introduction of heating may be expected to dry out timber components. Similarly gaps may be expected to open and close if heating is on during the day and off overnight.

Traditional materials and designs are, by definition, such as to accept the normal range of external and internal temperatures. Exceptionally high or low temperatures are indeed exceptional and rarely likely to feature as a significant factor in a building's deterioration. Perhaps the most common occurrence would be the very high temperatures achieved in a fire. The effect of

Figure 7.5 A mastic finished movement joint, John Radcliffe Hospital, Oxford.

Wear and tear

Often insurance policies will specifically exclude the provision of replacement components where the inadequacy is due to normal wear and tear. What is normal? What is wear and tear? It is deterioration in a component due to its use. For instance a carpet will be subject to wear and tear from people walking upon it. Normal wear and tear for a domestic-quality carpet will be that it would receive in a family house; not that of a hotel lobby which would enjoy (or endure) a much higher 'footfall', and where a higher, commercial-quality carpet would be appropriate.

Most components will receive some form of wear and tear – doors and windows opened and closed; taps similarly; lights switched on and off; lifts and stairs; kitchen fittings; furniture. Some of these lend themselves to testing and the development of standards. For instance, hinges can be specified according to the number of opening/closing movements they should withstand. Mild steel butts may be expected to be more durable than brass, and a pair-and-a-half to be more long-lasting than just a pair to support the same door, though this is largely due to the door being better restrained against distortion (which would cause uneven wear).

An owner or user of a building that is more intensively used should expect more wear and tear. At times of economic stress, with organisations seeking to 'sweat' their built assets, there may be pressure to accommodate more people in the same workspace, or to extend

Figure 7.6 Cracked and broken wall tiling, John Radcliffe Hospital, Oxford.

hours of use. 'Hot-desking' may be introduced to gain fuller utilisation of workstations. Such practices can easily increase utilisation, and thereby wear and tear, by as much as 50%. Thus a carpet expected to have say a 10-year life may be becoming threadbare after 6 or 7 years, or earlier.

Projecting from past performance

How long might a component be expected to last? What is a reasonable durability? How might one determine appropriate answers to such questions? Are there tests that can be done; what tests have been done? What are appropriate test conditions? Who might be trusted to provide information? Where to start?

Is it reasonable to assess how durable a component will be by ascertaining how durable similar components have been in the past, or how well they are performing at present? Past (and/or present) performance may be the best, or only, guide to the future; or it may be quite inappropriate. 'You can never plan the future by the past' (Edmund Burke, 1791).

What data is available? How complete is it; how reliable? For instance, steel casement windows in a building constructed in the 1930s are quite likely to be those built in originally; but they may not be. They could be replacements provided in the 1950s, 1960s, or subsequently. Records are often incomplete or non-existent. It can sometimes be worthwhile to track down a long-standing

Figure 7.7 A house in Woodstock, Oxfordshire.

resident or employee who may have a memory of such activities. That will still not tell you how long the windows may last now though.

Data vs. information

A window was replaced on 12th January 1998 – that is an item of data. Forty-eight windows, the entirety of those on the west elevation of the Bloggs Building, were replaced in the financial year 1997/8 as part of a programme of replacements across the Smith Estate – that is information. Further detail could be added – for example that the windows had been aluminium vertical sliders that had been subject to substantial and recurrent call-outs to respond to problems with spiral balances inhibiting satisfactory opening and closing. That the replacement windows were PVCu horizontal pivot windows, double-glazed, supplied and fixed by Jones & Co. of Oldcastle for the sum of £x would be further information.

A few years ago, seeking case study material in relation to so-called 'intelligent buildings', I visited an organisation that was strong in the market of environmental sensing and control technology. I was expecting to see a good example of application, of 'practising what they preach'. What I found was an organisation long on data but short on information. They had records of temperatures at various points in the various buildings, recorded at half-hourly intervals over a period of many months, over a year. However, there was no analysis of for how often, or for

Figure 7.8 A house in Longwall Street, Oxford.

how long, or by how much those temperatures were outside the designed (or acceptable) range; it was not part of anyone's job to perform such analysis. Nothing was done with or as a result of the data; it had great potential use but was unused. The data could have been turned into useful information driving actions to reduce heating and cooling loads and to increase user comfort – perhaps also staff retention and productivity. It is said that 'information is power' – certainly in this example such information could be used powerfully to effect changes for the better.

This is not to suggest, however, that volumes of data should be collected and huge processing power (whether of computing or technical personnel) allocated to its analysis and evaluation. The key is to devote appropriate technical resource to identifying what is worthwhile data to collect, and how it is to be collected, stored, analysed, reported and kept up to date. There is no point collecting data that will be unused or of limited value. It is also important that such data should be recorded consistently.

Limitations

There is much evidence of variability where surveyors make subjective, albeit professional, judgements on condition of elements and what should be done (Lomas, 1997; Hollis & Bright, 1999; Kempton *et al.*, 2002). Is it worth collecting data on that basis? Certainly it is worth trying to eliminate (or at least minimise) such variance. This could be done by use of just one surveyor (although that person may have 'good' and 'bad' days or be influenced by changing factors or context), although that may extend the data collection process. Team training, testing, double entry and normalising are further possibilities. Technology also offers opportunities. For instance, sensing devices can be used and calibrated to send alerts when certain criteria have been met or breached, to initiate and to record subsequent (perhaps automated) responses.

GIGO (garbage in, garbage out) is a recognised danger in assembling and assessing data. It reinforces the importance of collecting the right material and recording it correctly. Double entry of data is rarely worthwhile other than for system-critical matters – it doubles data entry costs, and may introduce anomalies to be resolved!

Data is also going out of date from the moment it is collected. Recording takes time. Buildings and their component parts are deteriorating all the time, albeit at variable rates. If constant rates of deterioration could be determined, it would be possible to build in a kind of depreciation rate and to anticipate when next inspections and/or interventions should be initiated. In the absence of such a model it is important to feed fresh data into the system at appropriate intervals. For instance, recognising the catastrophic effect of roof leaks on occupied spaces below, it makes sense to inspect roofs at least annually (especially flat roofs, which are more prone to failure). New data resulting from such inspection needs to be entered into the system. Similarly, when repair and/or replacement work is carried out, a record of that work needs to be made. It must be recognised that any data held, especially the more 'elderly' it is, needs to be checked out before action is taken based upon it.

Functional obsolescence

Not everything deteriorates because of use, or under the effect of weather. Components may be replaced for a range of reasons, including changing (increasing) standards, changing fashions and lack of availability of replacement parts or spares.

Functional obsolescence occurs when some part or component or the whole of a building fails to meet what is expected or to provide the performance that is required. For instance, kitchens in houses built in the 1930s and 1950s became functionally obsolete in the 1960s with the growth of the market for electrical appliances such as food mixers and blenders, refrigerators and washing machines, and subsequently electric carving knives, coffee grinders, espresso machines, dishwashers; also portable TVs, radios, CD/DVD players . . . The size of the room was inadequate and there were insufficient electrical points.

Offices become functionally obsolete when they no longer meet the commercial need. This may be manifested by an absence of such as air-conditioning or lifts, or an inability to accommodate growing IT needs or changing work practices. When what a building provides ceases to match what the occupying organisation needs, then analyses need to be undertaken and decisions made. For instance, could the building be adapted in some way to meet the need; how important is it for working practices to be changed; are premises available (hopefully nearby) that will better meet the new need? How readily might those new premises be adapted to meet anticipated future needs?

People will speak quite often of 'built-in obsolescence', meaning that a product has, built into it, something that means it will become dated and will need to be replaced at some stage in the future, perhaps quite soon. The suggestion is that the manufacturers and marketing men have a malign intention to feed desires for (simple and easily led) consumers to have the latest in fashionable and expendable gizmos. I have no wish to encourage such a low opinion of fellow men. An alternative way of looking at this is to talk of progress and technological advance.

In the 1950s and 1960s, people spoke of the domestic appliances referred to above as labour-saving devices. Our buildings have built-in obsolescence in the sense that over their lifetimes there is bound to be a significant amount of change that occurs in and to them, to which responses will be required. That is not that buildings are designed so that they will become out of date and so that expenditure will need to be incurred; what is required of them changes.

Some say that buildings should be designed to accommodate change; that they should be flexible or adaptable (e.g. Gordon, 1974; Groak, 1992; McGregor & Then, 1999). There was some discussion in Chapter 2 of the virtue, or otherwise, of such flexibility, and this is developed further in Chapter 16. It is very difficult to forecast directions, magnitudes and timescales of change. Who, when designing office buildings in the 1950s would have forecast (let alone designed to accommodate change to) Burolandschaft or open-plan offices? Who would have forecast the desktop personal computer or PC; or home working or hot-desking or WiFi?

In *Building Care* (Wood, 2003, pp. 152-169), I suggested that the difficulty of forecasting and designing for the future was such that it was perhaps better to limit aspirations to meeting the need of today, which is difficult enough.

A particularly topical issue at the start of the twenty-first century is the sustainability of buildings in the face of climate change. How and to what extent should our buildings be designed to accommodate to higher and/or lower temperatures, higher and lower rainfalls, higher winds, greater variability in weather and wider range of extremes? Will building owners and occupiers tolerate broader ranges of internal environmental conditions; or will they be more demanding of consistent and high performance?

Perhaps a modular approach to building components will facilitate easier repair and replacement of components, and periodic upgrades as replacement components are made available that will achieve higher standards of performance. This may enable functional obsolescence to be resisted and redressed. For instance, it may be possible to replace a fully glazed window panel with one with a smaller glazed area, or one which provides higher thermal insulation. It will still be necessary of course to evaluate whether it makes sense to make a piecemeal upgrade or to initiate a longer-term or larger-scale programme or to replace on a like-for-like basis or any other action. Whether it makes sense to adopt the modular approach just in order to facilitate easy upgrades that may or may not be carried out is another issue.

In short, functional obsolescence is almost certain to occur. We cannot be certain about just what will be the elements that become out of date soonest or to the greatest extent of unacceptability. Perhaps the best, or least, that should be done in anticipation is to build up either financial reserves or sufficient creditworthiness as to be able to make the necessary investment in upgrading the building, or moving, when the time comes.

Case study: a local authority

Local authorities, like public sector bodies as a whole, have suffered from many years of under-investment, manifested in deteriorating buildings, and large and increasing maintenance backlogs. This authority resolved that to better direct resources into the future it should adopt a more proactive role.

Background

The authority had been formed from what had been seven former authorities, a mixture of rural and urban district councils, with around 7000 council-owned dwellings to be managed and maintained. The properties were located in and on the periphery of the four major towns and across the 180 square miles (470 sq km) with a pair of semi-detached houses or a short terrace in just about every village or hamlet.

The age profile of the housing stock was fairly typical of a non-metropolitan local authority:

■ 20% inter-war, typically semi-detached, with brick, rendered or half-rendered elevations beneath a slate or clay-tiled roof;
■ 50% immediate post-war (1950s), with a small number of prefab bungalows, a number of prefabricated reinforced concrete (PRC) houses and flats, and the majority two-storey houses in short terraces, traditionally-built, brick and tile;
■ 20% 1960s, typically three-storey blocks of flats with large timber windows and flat roofs; no tower blocks;
■ 10% 1970s and since, looking generally more like typical private estate houses, a jumble of short terraces with a lot of staggers, timber boarding and tile hanging.

The stock was managed through one central office with three maintenance depots. Some plans existed of some of the developments, largely with imperial measurements from the time of their construction; most of these were rolled up in basement archives, of historical interest but little value in practice.

Issues

The buildings displayed defects typical of their age and design. The council was conscious of its obligations to keep their housing in a good state of repair; maintenance needs were increasing and there was a growing backlog of work. Tenants were also developing higher expectations of service from the council. Sales of dwellings under the government's right to buy (RTB) legislation were also an increasingly complicating factor.

To an extent the organisation was based around inherited systems and priorities without a unifying strategy. It was appropriate to 'take stock', in several ways, in order to plot a way ahead that would better meet the needs and 'deliver the goods'.

Maintenance was broadly carried out in response to repair requests. Technical officers would order up work based on their prior knowledge of the properties on their 'patch', visiting as much as was necessary to fill gaps in knowledge and to check up on work in progress or completed as and when time permitted. Much reliance was placed upon this corporate memory. A small number of policies existed to update some elements, for instance to replace broken cast iron or asbestos–cement rainwater goods and rotten timber or rusted metal windows with PVCu on the understanding that these would reduce future maintenance needs.

Action

It was resolved to carry out a stock condition survey. Consultants were engaged to design and execute the survey as the authority's own staff were fully engaged in their current work, which needed to continue.

■ A 100% external survey was carried out of all dwellings, and approximately 15% were inspected internally – one of each major plan type.
■ Pro formas were prepared (developed from a previous study elsewhere), pilot tested for ease and accuracy of completion and agreed.

■ Photographs were taken of each believed plan type and variant in each location and compared in the office to identify like designs and allocate type codings accordingly.
■ Dwellings were measured and floor plans were produced to consistent style and scale (1:100) to inform future maintenance and upgrading.
■ Location plans were also produced for each estate or grouping, with plan types coded and marked up.
■ A 10-year programme of elemental repair and improvements was developed.

The survey exercise took about 18 months, consisting broadly of:

■ a 3-month preparatory period including discussions between client and consultants to finalise the brief, agree survey documentation and outputs;
■ 12 months of survey field work with a team of six (a mixture of about 50:50 experienced from 'HQ' and recruited locally for the project) collecting, recording and inputting data, with data analysis being carried out in parallel;
■ 3 months beyond completion of the last survey to complete analysis, make overall assessments, prepare and present proposed programme for approval.

Findings

The comprehensive and coherent collection of data identified a level of maintenance need substantially greater than that previously being met. Care was needed to allay fears that this may be construed as officers not being 'up to the job' but a true summation of the work required and not previously assembled in this way.

There were consistent patterns of failure that could be identified for dwellings of particular designs or construction and that some parts of the district were more in need of attention sooner than others. This enabled officers to resist, or at least respond to, the blandishments of individual councillors for a greater share of 'the pie' than warranted by looking at relative priorities overall.

A lot of asbestos–cement rainwater goods were identified, enabling the programme to work toward their complete removal; and the programme of window replacements was substantial enough to warrant the authority investing in its own assembly line for the production of PVCu units.

Reflection

Concern was expressed at the time of appointment about the quantity of data to be collected; this would need to be stored and regularly updated if it was to retain its currency and value. Officers felt that it was worth collecting a substantial amount of detail 'while you're there'. Data was collected on the colour and texture of bricks. That was difficult enough as many bricks are a kind of orange reddish brown mix and their hue depends very much on the weather and how much rain they have recently experienced. Consistency of surveyor assessment and description were difficult to achieve. When considering that work is carried out on brickwork infrequently and that its specification would almost certainly involve a visit to make a technical assessment, collecting this data was of limited value.

Similarly, some of the plan data was of limited value at that time. A more targeted or phased approach, focused on the more immediate needs initially could have produced quicker 'wins' and buy-in from seeing action on the properties (rather than on the survey). A survey programme that built in more involvement of in-house staff would have not only obtained better input and buy-in from them but enabled a fuller integration of the survey data into an ongoing maintenance management tool.

Perhaps the most cost-effective part of the exercise was the photographic and type coding component; it gave everyone involved an instant picture that would conjure responses along the lines of 'Ah yes; I know the problems with that type'. The project substantiated the scale of repair and improvement needed and supported the case for more generous and consistent investment in the properties and their effective maintenance.

Summary

This chapter has looked at the deterioration and decay of buildings and services over time. It has reviewed a range of climatic and weather-related factors that degrade the external elements of buildings. It also considered briefly some of the attributes of building materials that contribute to durability or decay, and showed that these are not only physical but also include economic and other factors. The key aspect is that buildings need to continue to function effectively; action is required when they don't.

In the following chapter some attention is given to ways in which buildings may be assisted to provide the required functions more closely for longer.

References

British Standards Institution (1992) *BS 7543: 2003: Guide to Durability of Buildings and Building Elements, Products and Components*. BSI, Milton Keynes.

Building Research Establishment (1989) *Thermal Insulation: Avoiding Risks* (1st edn). BRE, Watford.

Burke, E. (1791) *Letter to a Member of the National Assembly*. In: *Oxford Dictionary of Quotations* (4th edn) (1992). Oxford Universtiy Press, Oxford, p. 157.

Douglas, J. & Ransom, B. (2007) *Understanding Building Failures* (3rd edn). Taylor & Francis, Basingstoke.

Gordon, A. (1974) Architects and resource conservation. *RIBA Journal*, **81**(1), 9–12.

Groak, S. (1992) *The Idea of Building: Thought and Action in the Design and Production of Buildings*. E. & F.N. Spon, London.

Hollis, M. & Bright, K. (1999) Surveying the surveyors. *Structural Survey*, **17**(2), 65–73.

Johnson, S. (1758) *The Idler, no. 11*, 24 June 1758. In: *Oxford Dictionary of Quotations* (4th edn) (1992). Oxford Universtiy Press, Oxford, p. 368.

Kempton, J., Alani, A. & Chapman, K. (2002) Surveyor variability in educational stock surveys – a lens model study. *Facilities*, **20**(5/6), 190–197.

Lomas, D.W. (1997) Team inspections of high-rise in Hong Kong and the UK. *Structural Survey*, **15**(4), 162–165.

McGregor, W. & Then, D.S.S. (1999) *Facilities Management and the Business of Space*. Arnold, London, pp. 26–33.

Richardson, B.A. (2000) *Defects and Deterioration in Buildings* (2nd edn). E.& F.N. Spon, London.

Watt, D. (2007) *Building Pathology: Principles and Practice* (2nd edn). Blackwell Publishing, Oxford.

Wood, B.R. (2003) *Building Care*. Blackwell Publishing, Oxford.

8 Building defects and avoidance

The previous chapter considered the deterioration and decay of buildings and services over time. It looked at how climatic and weather-related factors degrade the external elements of buildings. It also considered briefly some of the attributes of building materials that contribute to durability or decay, and showed that these are not only physical but also include economic and other factors. The key aspect was that buildings need to continue to function effectively, and consequently action is required when they don't.

This chapter focuses on how buildings may be assisted to provide the required functions more closely for longer.

BS 4778-2: 1991 *Quality Vocabulary – Part 2 Quality Concepts and Related Definitions* (British Standards Institution (BSI), 1991) identifies 'defect' as one of two words (the other being 'item') that are 'terms which merit special note'; it advises thus:

> *2.2.1. Defect*
> *Care should be taken in the use of this term. In the United Kingdom the term is frequently used for a product or service that is found to be unfit for its intended usage. Some national and international glossaries relate defect to a condition that renders an item unsafe or incapable of meeting functional or other customer expectation.*
>
> *The IEC {International Electrotechnical Commission}concept of a defect relates to all conditions that can give rise to losses and product liability. Legislation varies widely from one authority to another.*
>
> *It appears likely that the definition of defect will remain under review for some considerable time and caution should be exercised in the use of this term.*
>
> *The term should not be used as a general term and it would be prudent to avoid its use altogether if possible.*

Hollis (2006, p.26) defines a defect by reference to the Norwegian Building Research Institute: 'unexpected expenditure incurred by the client following taking possession of a property', adding his observation that 'the defect is determined by the client and not by a building professional, and it is identified by unforeseen expenditure'.

It is very difficult to write about matters in buildings which are not as they should be, without recourse to the term 'defect'. It is recognised that one person's defect may not be so to another, and that is an issue to be addressed. It is also important to recognise that the nature and/or possible causes of a defect are also very relevant to its consideration, to its rectification and to its potential avoidance. I shall therefore endeavour to discuss defects in a context, describing the nature of the defect rather than to just refer to an item as 'defective'. Unless the context of the word is such as to suggest some other meaning, the term 'defect' should be understood as to relate to a building or some part or service that is considered by someone to be in some way inadequate for its situation.

Avoidance of defects

The best way to deal with defects and shortcomings in buildings is to avoid them in the first place; this is almost invariably the cheapest approach too of dealing with defects. Avoidance is very different from evasion – as any advisor on taxation will tell you. To avoid something is to take positive steps to try not get into a particular situation by taking steps to ensure the situation does not arise; to evade is to fail to accept one's obligations, to ignore, to fail to deal adequately with a situation that has arisen. In the case of taxation, avoidance is acceptable (and personally desirable); evasion is illegal and can lead to gaol.

Almost certainly, once defects have arisen they should not be evaded; they must be considered and dealt with appropriately. Usually the most appropriate way to deal with a defect will be to rectify it. Exceptionally, for instance if the building is due to be demolished shortly, it may make sense to live with a defect for a time – the decision, however, is not evaded. The situation has been considered and the way it is to be dealt with has been decided.

The UK Building Research Establishment (BRE), working in conjunction with the then National Building Agency (NBA), carried out a 3-year study of traditional housing to identify common problems with a view to reducing their occurrence in the future; I was involved with that project. The investigation included scrutiny of drawings and specifications and inspections of construction in progress on site. The study showed that there were broadly similar numbers of faults attributable to poor design and/or specification as to poor work on site, and that the cost of that faulty design or work and its remedying was substantially more than to have done it correctly. The report produced advice and checklists related to design and construction stages, aimed at avoidance of faulty construction; it defined a fault as:

> a departure from good practice as defined by criteria in Building Regulations, British Standards and Codes, the published recommendations of recognised authoritative bodies and (for faults of site origin) a departure from design requirements where these were not themselves at fault. (BRE, 1982)

How can defects (faults) be avoided? Defects can be created from the outset, even before the building project is formulated, at the design stage, or during construction or during the life of the building in use; these four stages will be considered in turn.

Defects at the inception/pre-design stage

The maintenance manager has a substantial potential for assisting in reducing maintenance (and thereby costs) by having involvement in all decisions about buildings, especially decisions to procure new buildings. It may be, for instance, that the maintenance manager could save the entire maintenance costs of a building by questioning the need for the building at all! What buildings are required for the proposed operations? Why this building or these buildings? Why this particular upgrade now? It is good to adopt a questioning attitude at this stage. Unfortunately this will not always make you popular as some people, for whatever reason, become committed to a 'pet project' and take questioning, however well intentioned it may be, as criticism, and take it personally. So it is worth thinking twice about what you wish to say and how and when may be the best approach. Think: what is the desired end result of your questioning? People can become quite defensive when put 'on the spot'. Hone up and then deploy your inter-personal skills.

Every new building or major refurbishment or refitting project involves substantial investment and commitment of time and money. As the saying goes, it is a 'shame to spoil the ship for a pennyworth of tar' – which is a way of saying that sufficient funds should be made available for what is really required rather than to try to do the job for less. However it is also possible to spend more time or money than is really warranted, and thereby deprive other projects or processes of resources that might have been better allocated to them. It is worth taking time at the earliest stage to consider and determine just what is the appropriate amount of resource (of time as well as money) to allocate to the project.

This could be done by any of a range of financial appraisals, such as:

■ return on investment (ROI);
■ return on capital employed (RCE);
■ net present value (NPV);
■ discounted cash flow (DCF);
■ life cycle costing (LCC).

These methods are not discussed here; there are many books available (e.g. Jaggar *et al.*, 2002; Flanagan & Jewell, 2005; Kirkham, 2007). Additionally or alternatively, simple payback or break-even calculations and sensitivity analyses can be done to help decide the most appropriate budget. There are so many variables involved, however, that decisions will often have to be made based on the relative weights one gives to each of the factors involved. Such discussions are well worth having. If the right questions are not asked, and answered, at this stage, we ought not to be surprised if the building fails to meet the true need.

The result of these deliberations should be a brief for the proposed built end-product. The brief should set out what is to be provided, or better still what is to be achieved or facilitated; it may include information on floor areas required, adjacencies, standards of performance to be achieved, quality of materials and so on. Many buildings, and reports produced relating to the difficulties of their construction, testify to the lack of an adequate brief. It ought not to come as a total surprise, perhaps, that so many building projects over-run in terms of their completion date and cost. Virtually every new building is to all intents a prototype, and refurbishments take place in buildings that are often not built as supposed and that may also be occupied. Such issues make it more imperative that briefs should be prepared that spell out as clearly as possible at the outset what the desired outcome is; this could also include indications of where flexibility of end-product, and/or in ways of attaining it, may be acceptable. The former suggests a *prescriptive* brief; the latter a *performance*-related brief.

Defects at the design stage

Many of the 'defects' created in the design stage will have their roots in the pre-design stage, that may or may not have been effectively undertaken. If the client was unclear as to what he or she or they wanted, then it should be little surprise if that is not what the design sets out to achieve. At risk of upsetting some of my friends and colleagues, architects are often drawn into that profession by a deep desire to do their part to change the world. Each design commission provides a vehicle and opportunity to contribute to that transformation. This can present temptation to perhaps 'overegg' the design in pursuit of a maybe more memorable built legacy than strictly warranted.

Architects, structural engineers, building services engineers and others involved in creating and refining the design have the responsibility of identifying, sorting, sifting and integrating relevant information – regulations, standards, recommendations, advice – from a wide variety of sources of various kinds. The design team has to share ideas and help devise the design, weighing the various considerations. Each of the participants must be expected to bring to bear their own experiences (good and bad) to the table together with their concerns, preferences and prejudices. It tends to help create more harmonious working relationships when participants have worked successfully together on projects previously, although this can provide a kind of cosiness that may result in ready consensus and unquestioning acceptance of a way ahead that ignores the potential for exciting innovation.

I have already referred to the BRE/NBA study of quality in traditional housing that found that many faults were created at the design stage. That study had stemmed from an earlier survey (Burt, 1978) that concluded that 'very substantial sums of money' were being spent 'putting right faults that should not have occurred'. As part of a follow-up exercise, BRE published through the 1980s a series of Defect Action Sheets, some related to design, others to construction. These covered common defects and identified errors and how the job should have been done properly. However, despite these good intentions, defective designs and constructions have continued. Egan (1998) estimated that £1 billion was spent in correcting construction defects. Atkinson (1998) found that there was generally no shortage of knowledge of how buildings should be built:

> *managerial influences underlie many errors leading to defects {and}... as a consequence, the continual emphasis placed by technical publications on correct technical solutions . . . is misplaced.*

Figures 8.1 to 8.4 illustrate defects that could have been identified and eliminated at the design stage. Figure 8.1 shows a rainwater downpipe passing through the mosaic clad concrete floor projection. It was probably unnecessary for the floor to project beyond the external walls at all (unless they were intended to provide shading to the windows beneath, in which case a lighter *brise soleil* could have been installed) or for the downpipes to pass through. This point of weakness has resulted in bad staining and damage to the concrete soffite.

Figure 8.2 shows staining to the absorbent concrete blockwork below the windows. Rainwater could have been expected to be shed by the window glazing, yet no cill has been provided to cast the water away from the wall. In Figure 8.3 the eaves detail fails to direct the rainwater into the gutter and downpipe (which appears to be very flat or even falling back), discharging it over the verge of the lower roof and soaking the brickwork below. The glazed roof shown in Figure 8.4 looks 'interesting' in terms of its maintenance.

So how are defects to be avoided at the design stage? It would seem that important components would include:

▪ deep knowledge on the part of (all?) members of the design team;
▪ experience (both good and bad could be put to good use);
▪ understanding of the context and of what is required;
▪ commitment and care on the part of all participants;
▪ clear criteria for assessment of the design;
▪ sound application of principles of good design;
▪ effective communication;
▪ simplicity.

Figure 8.1 Design detail: leaking where downpipe passes through structure.

Shortages in some of those can be overcome; for me the essential prerequisites for success would be:

■ avoidance of complexity;
■ maintenance of commitment;
■ honesty in communication.

Apply the KIS principle: keep it simple.

Remember that no-one owes you a job; someone is paying for the building. Real, feeling people will use the building – all these people depend on the design team.

Mark McCormack, the sports entrepreneur, identified three 'hard to say' things. Use them often and encourage all in the team likewise:

■ I don't know.
■ I need help.
■ I was wrong.

Denial in any of these areas leads to error.

As you will, I hope, appreciate, all these are people-related, management rather than technical issues.

Figure 8.2   Design detail: rainwater shed from windows on to blockwork beneath.

Defects occurring during the construction stage

Many of the same issues apply as at the design stage, but with the added complexity of having many inter-related activities happening in a constricted space against perhaps unrealistic deadlines, and with many people of disparate backgrounds involved for the first and perhaps last time.

Many of the defects will arise from the design. It is worth the contractor taking time as early as possible to scrutinise the information of various kinds provided – drawings, specifications and other contract requirements, standards, suppliers and their recommendations, working constrictions and so on – to identify potential issues as far as possible. It will then be possible to at least seek resolution of likely problems before they occur. Even though these shortcomings in the design may be the fault of the design team, it is better to concentrate on getting them resolved rather than looking to lay blame. This will be a lot cheaper than having to remove and replace defective work, let alone going to court to battle out who is at fault.

What then are the likely causes or sources of defects during construction? Faulty materials or components could be supplied and/or incorporated. The possibility of incorporating poor construction work should be reduced by carrying out periodic and routine sampling and testing for compliance against appropriate criteria. I say *appropriate* criteria because it is possible to comply with a British Standard or with Building Regulations without the construction being very good – those should be considered as absolute *minimum* standards. Evidence suggests however that it is often difficult to obtain even these lowest of standards.

Figure 8.3 Design detail: rainwater shed from eaves/gutter on to brickwork beneath.

What is an acceptable standard? BS 3921: 1985 permitted a dimensional tolerance of ±75 mm on a length of 24 bricks laid end to end. While that is a difference in length of around ±1.5%, which may not sound much, the difference between one brick and another can be quite large. Brickwork is designed to have mortar joints of 10 mm, but with these tolerances one could be faced with brickwork with joints varying between 6 mm and 14 mm – a huge and very noticeable difference. To demand closer tolerances requires either more consistency in manufacture or greater rejection rates, both of which add expense and scope for disagreement as to acceptability at the margin. There is scope for disagreements on variations in shape, colour and texture too. There is much to be said for specifying new work to contrast rather than blend with existing – matching is always difficult.

BS 3921 was withdrawn in 2007 and superseded by BS EN 771-1: 2003 which permits a range of tolerances to be defined.

Returning now to components more generally, they can also be damaged in a number of ways:

- in transit to the site;
- in transfer to storage on arrival;
- in storage;
- in placing in position in the building;
- while in position and by work in progress before completion.

Figure 8.4   Design detail: glazed roof.

Care should be taken to protect components, especially faces, edges and corners.

The building should not be accepted as complete, even practically complete, with defects. Some will say that that is a counsel of perfection (meaning unachievable); I say that that is what the client is paying the contractor for – compliance with the contract between them.

What is meant by 'practically complete'? To paraphrase Humpty Dumpty perhaps: 'It means whatever the architect wants it to mean' (Lewis Carroll, Through the Looking Glass). The term is not defined in the Joint Contracts Tribunal (JCT) forms of contract. Court cases have given rise to differing understandings. I do not intend to develop the arguments here other than to encourage you to develop good and harmonious working arrangements and thereby to avoid litigation. Interesting discussions can be found on two legal websites (www.alway-associates and www.rics.org).

However, the pressure is on (whether or not explicitly) to accept the building as complete when it really is not. Most building projects finish late and clients are keen to either move in or get tenants in as soon as possible in order to move from spending money on the building to earning money from it! Quality tends to suffer in the quest for completion. Clients and their architects and clerks of works are said to be unreasonable and unhelpful when holding out for the correct standard; undertakings will be given to correct non-conforming work during the defects liability period.

The defects liability period is the time after practical completion of the building during which the contractor retains responsibility for correcting defects which have become apparent during that time. Under the JCT forms of contract it is normally 6 months, although 12 months is recommended for services installations (so that they have been submitted to the whole of an annual cycle of use of heating and cooling systems). Normally, these defects will be things like small shrinkage cracks between ceiling and wall plaster, but often includes defects which should have been made good prior to completion, such as scratched and damaged fittings, finishes and glazing, and poor decoration. The correcting of these defects while the building is occupied can often be extremely inconvenient and disruptive to operations, as well as being undesirable.

I encourage all to establish and hold out for appropriate standards at and from the outset. It is sometimes said that contractors price projects according to the weight of documentation provided. I contend that the greater the detail provided, the greater the clarity of just what is required (subject of course to there being no duplication or conflict in information!).

Attitude and training are the keys to success. The architect or contract administrator or supervising officer must be able to trust the contractor to build correctly, and the contractor must instil and uphold that same attitude in the tradespeople. Temptations to cut corners must be resisted. Clients and contractors should be encouraged to insist on a fully qualified workforce. The Construction Skills Certification Scheme (CSCS) was introduced to help instil customer confidence and improve the quality of construction, and thereby eliminate the 'cowboy builder'. CSCS cards list the holder's qualifications and assure health and safety awareness. As at 1st August 2008, 1 296 883 CSCS cards had been issued. Clients and contractors need to check that operatives have the requisite cards.

Defects while the building is in occupation

As was explored in the previous chapter, buildings and their component parts are deteriorating from the time they are produced, albeit slowly we hope. The rate of fall off of performance may vary over time. External factors may get worse or they may abate. For instance, it may be thought that climate change will increase and exacerbate deterioration. In the opposite direction, the Clean Air Act of 1956 helped produce a less aggressive environment. In other circumstances the rate of deterioration of a component may be more a factor of internal decay 'built in', such as timber.

There may be changes in use or changes in intensity of use of the building that may have injurious effects that manifest themselves in defects – elements that cease to function or that function less well sooner than might have been expected.

Buildings or component parts may be subject to misuse, to misunderstanding of how they should be used or of how they were designed to be used. They may suffer from maltreatment or vandalism. Perhaps there may be elements of the provision that were insufficiently robust for their situation.

There are also defects that may occur as a result of wrong or inadequate or overzealous cleaning or other maintenance regimes. The provision of manuals at hand-over on completion of the building, setting out how it should be looked after is still not common – it should be a regular expectation.

Perhaps the provision of energy certificates at hand-over, showing the expected energy performance of the building, and energy performance certificates showing the actual energy performance of the building in use, will encourage the development of both more and better documentation at hand-over and greater expectation of performance review through the building's life.

Similarly, growth in interest in operational costs of buildings funded through the Private Finance Initiative and in whole-life costs generally may prompt a greater interest in studying and projecting profiles of likely defects as buildings deteriorate and expectations grow and change. Anticipating defects may be a first step to avoiding them.

How to avoid defects

Having considered broadly how defects may be created (and subsequent chapters will investigate defects in more detail) it is important to put focus on the positives of how to avoid the creation of defects. Figure 8.5 shows a simple step-by-step approach to the avoidance of defects.

'Zero defects'

How can zero defects be achieved? Is it achievable? 'It's impossible' may be your view. Why should it not be possible? Is it that we believe the construction industry is unable, or unwilling, to meet the challenge. Why is it not an everyday expectation and achievement?

The British motor industry was transformed in the 1980s by the introduction of Japanese management techniques and standards. My previous book *Building Care* has a whole chapter devoted to the Japanese influence on thought and practice as applied to industry (Wood, 2003, pp. 49–67). Japanese industry learned a lot from that of the USA in the years of reconstruction after World War II, and became a significant global competitor in the 1970s and 1980s. The Japanese put a lot of stress on attention to quality. Cole (1999) reported that Yokogawa Hewlett-Packard, a Japanese/American joint venture slashed the 'wave solder rate of non-conformity' from 4000 parts per million (PPM) in 1977 to 40 ppm in 1979 and to 3 ppm in 1982.

Mozer (1984) identified six components of total quality control:

1. **Commitment** to continuous quality improvement, led by top management.
2. **Collection** of data.
3. **Clarity** of who is responsible for day-to-day decision making on problems.
4. **Customer feedback** gathered systematically.
5. **Use** of the **Deming cycle** (plan–do–check–act–plan, do. . . etc.) as a problem-solving process to achieve permanent solutions.
6. **Use** of **statistics** as a management tool.

Implicitly, every error has within it a germ, the potential, of its avoidance in future. Every time some unacceptable product is produced it must be analysed to ascertain the cause so that any necessary changes can be made to processes for the future. Now it may be argued that such processes may be more appropriate and applicable in an industrial environment; 'the construction industry is different'. Certainly construction has a lot of seemingly one-off situations, especially in refurbishment projects, but to focus on the differences is to avoid the learning possible from looking at the similarities. This analytical process will be applied in more detail in Chapters 10–15 to the defects discussed. An example is considered here.

Let us consider a new door hung in an existing frame *in situ* in an existing, occupied, building: the door doesn't close correctly.

1. **Commitment.** The door doesn't close correctly – that is unacceptable. It is no good suggesting that it's OK or that it can be lived with; it is not OK. No-one in the client's or contractor's organisation should accept it.

| 1 | START – Get into the right frame of mind (How?) |

| 2 | Decide whether you expect to **reduce** or **eliminate** defects |

| 3 | Adopt this checklist (or adapt it) or decide to develop some other approach |

| 4 | Ensure that the brief does not **encourage** the creation of defects, e.g.
 Engage in full discussion with the client to ascertain the true **need**
 Ensure that there is enough **time** for the design development
 Ensure that **budgets** have been determined that reflect the **quality** required by client |

| 5 | Collect and evaluate **data** on similar buildings in use |

| 6 | As far as possible engage contractor fully – design and development |

| 7 | As far as possible use tried and tested materials and contractors |

| 8 | During design development check designs (including specification) **weekly** for:
 • weather tightness
 • buildability
 • design quality
 • durability
 • maintainablitiy
 • other? |

| 9 | Obtain formal **sign-off** from experienced 'critical friend' |

| 10 | Insist on fully qualified workforce on site (names and qualifications to be provided) |

| 11 | Check work on site **daily**, not for conformity with contract but for **fitness-for-purpose**:
 • weather tightness
 • joints
 • finish
 • durability
 • maintainablitiy
 • other? |

| 12 | Obtain sign-off from nominated representative of contractor |

| 13 | Receive full documentation on maintenance and cleaning requirements |

| 14 | DO NOT accept handover without satisfactory **completion** |

Figure 8.5 Defect avoidance routine: a checklist.

2. Collect data. What is it that is wrong so that the door does not close? Possibilities include:
   ■ The door is too tall; or too wide; or too short; or too thick.
   ■ The wrong dimension of door was ordered.
   ■ The wrong door was delivered.
   ■ The wrong door was fitted.
   ■ The door is distorted: bowed, cupped, twisted.
   ■ The door was poorly stored, protected, delivered.

- The door was damaged.
- It was wrongly hung: too few hinges, wrong size, wrong or insufficient screws.
- Too many coats of paint.
- Wrong ironmongery.
- Frame should have been replaced (too).
- Frame is distorted.
- Frame has moved between measurement and action; has come loose, twisted.
- Building has moved, deflected.
- Floor or ceiling is not horizontal.
- Frame is not vertical.
- No allowance made for (moisture) movement; drying shrinkage, swelling.
- No allowance for threshold or carpet or doormat.
- Other; none of the above.
- Combinations of the above.

3. Clarity of who is responsible for day-to-day decision-making on problems. Several people are likely to have been involved in the process from selection of what door and/or what (other) action was needed through to its fitting and subsequent use. (It may be, for instance, that the door did close when fitted but has been damaged subsequently.) Possible questions in relation to this aspect might include such as:
   - Who decided a new door was required?
   - Who measured up for the new door?
   - Who ordered it? Similarly its ironmongery.
   - Who dispatched the door?
   - Who received (and accepted) it?
   - Who arranged its storage and protection?
   - Who fitted the door; was it regarded as OK – if so, by whom (and why)?
   - Who has reported the door as not closing?
   - Who should have identified the defect before it was accepted/handed over?
   - Who is responsible?

4. Customer feedback gathered systematically:
   - Is this the first time such a defect has occurred (or been reported)?
   - If not, how often; how long ago; what happened?
   - Is this in line with previous performance; or worse/much worse?
   - How content is the customer?
   - What undertaking might be given for the future?
   - How and when will customer satisfaction be checked?

5. Use of the Deming cycle (plan–do–check–act–plan, do... etc.) as a problem-solving process to achieve permanent solutions.
   - What processes were in place regarding decision making; measurement; ordering?
   - What shortcomings were there in those?
   - What changes could be made, e.g. suppliers, staffing, staff selection and qualification, checking, training, customer care, attitudes?
   - Decide on changes to be made, and checks to be made (by whom and when) to ascertain their effectiveness.
   - Effect requisite changes.

6. Use of statistics as a management tool:
   - Inspect regularly and irregularly.
   - Prepare regular *exception* reports (appropriate period to be determined) – i.e focus on what is *unusual*.

- ▪ Require contractors to report on defects to client.
- ▪ Set targets for next period based on (halving?) number/type of defects of the previous period.
- ▪ Recognise (and reward?) good performance.

What is crucial is that all involved should care. One of the key principles in relation to quality is that the person doing the job should be concerned and responsible for doing it right. If they don't know what they should be doing, or how, they should say so, seek advice and act accordingly. If they get it wrong they should expect to put it right or have someone else do so, perhaps at their expense. Any system, however, that encourages people to cut corners and hope to get away with it is doomed to defects.

Repair/replace decisions

There is potentially quite a lot of expenditure at stake in the making of decisions about whether a particular situation warrants a repair or a replacement. There is also expense (in the form of professional time) involved in the making of the decision. Time is also important – it will take time to get the professional involved and for him or her to make the decision; also to have the decision implemented.

One way of reducing that professional time, and reducing the total time to get the repair or replacement effected, is to develop a set of criteria for decision making. This will also reduce the scope for variation, and improve certainty and consistency in these decisions. It is important, however, not to introduce a kind of 'automaticity' of response that cuts out the possibility of professional input and over-ride that may be warranted by the situation that is actually not as standard as had been envisaged.

Figure 8.6 represents an attempt at drafting a generic repair/replace decision-making tool that may be useful in forming an overview to inform local responses. Whatever pathways are taken to delivering the repair or replacement decided upon, it is important to be methodical in the taking of that decision and in the making of records of such decisions. In the example in Figure 8.2, the first level of consideration and decision is whether to repair or replace the component, element or building. Firm criteria against which to decide will help. If the decision is to replace, then a further consideration must be made and decision taken as to whether the element should be replaced on a like-for-like basis or with something different. The replacement should be monitored and its performance – whether, for instance, better than previous or a disappointing outcome – should be recorded in a database and used to inform future decisions. Similarly considerations should be made for repair decisions, for instance to apply a quick fix or to make a more thorough repair. The process allows for effective decision-making processes to be developed and thereby good decisions to be made and put into effect.

No decision is a straightforward, black-and-white, cut-and-dried matter; there are nuances, shades of grey, matters of context to be taken into account. For instance, if funding is in short supply – the usual situation – a repair will be the normal expectation. If, however, repair bills are mounting, that will provide impetus for a wider or longer-term overview to be taken and a programme to be prepared that is more likely to include at least a proportion of replacement works to be instigated. The organisation's longer-term objectives and commitment to the particular building will also be considerations. If a building is thought to have only a short or uncertain life ahead, replacements are only likely to be warranted if required by the lease or in order to facilitate sale or letting.

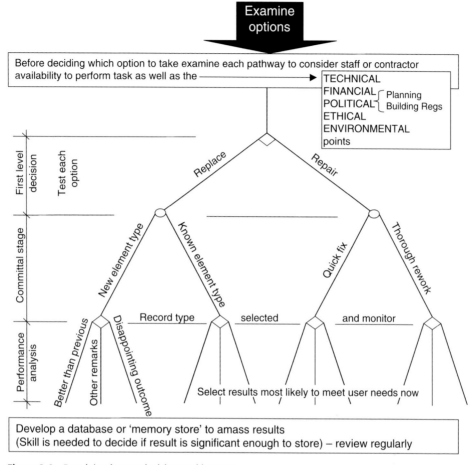

Figure 8.6 Repair/replace: a decision-making tree.

The amount of time devoted to decision making on repair matters, and the urgency accorded it, should be proportionate to the potential impact of the decision rather than the cost of the repair itself, which is likely to be small in the overall scheme of things. Thus, decisions related to matters of health and/or safety should be made promptly but not hastily, taking into account all relevant matters, with cost being low in the order of consideration. That is not to say that cost is immaterial when there are choices to be made, but that life is of infinitely more value.

Specifications

This is an area often somewhat overlooked or downplayed in new building work, where the drawing is the key medium of communication. It is said that 'a picture is worth a thousand words', and there is a lot in that; however not everyone is fluent in that language. Perhaps a picture tells a thousand stories, especially if one considers that 'beauty is in the eye of the beholder'. Pictures may be read differently by different individuals. Often the most useful drawing is the detail not provided of just how the various components come together at a three-dimensional junction. One of the suggestions arising from the BRE/NBA study (BRE, 1982, p.20) was that 'three-dimensional drawings should be produced'. Can you picture Escher's three-dimensional drawing of a staircase – it keeps going round and round and up and up in a continuous loop; if it is possible to draw it, surely it must be possible to build it? It is not good for architects to give constructors such challenges. Indeed it may be that a *four*-dimensional approach should be adopted, focused not so much on what the finished product looks like but what goes where, and when – which parts go in first? This would be rather like the exploded views shown for creating a model aircraft from a set of plastic parts in a box, or a self-assembly piece of furniture.

Certainly, defects occur commonly at junctions and intersections, so focus should be put upon those locations. The 'Common Arrangement' produced by then Coordinating Committee for Project Information (CCPI) (later the Construction Project Information Committee – CPIC) (CCPI, 1998; CPIC, 2003) was intended to give a common approach to how design and construction information would be communicated, so that buildings may be better constructed. Constructional drawings would be of a small range of kinds dependent upon their purpose – location, assembly or component.

Location drawings would, for instance, show where the building was located on the site, give overall dimensions, be fairly general and short of detail, and indicate the points of structure (normally intersections) which would be shown at larger scale and in greater detail elsewhere. Location drawings would typically be at scales of 1:200, 1:100 or 1:50; site location plans at 1:1250 or 1:2500 to accord with scales used by the Ordnance Survey for mapping of (respectively) urban and rural areas of Britain.

Assembly drawings show the intersections of parts of the building. An eaves detail, showing the intersection of roof and wall would be an example. The most common and useful scales would be 1:5 and full-size (half-scale is not recommended as, despite clear statements that drawings should not be measured from, and only stated dimensions to be used, these are too close to the appearance of full-size and therefore confusing.)

Component drawings show features such as windows, doors and the frames into which they should fit, and panels of various kinds. Typical scales would be 1:10 or 1:20. They may be fairly sketchy if the component is a standard one, for instance from a manufacturer's catalogue, or in detail if to be specially made from this drawing. They may incorporate, or be accompanied by, schedules of quantities of each component. Their locations in the building will be shown on the location drawings, and how they fit together on an assembly drawing.

This kind of drawn approach is very appropriate for newbuild, especially where there is a degree of repetition, either within the one building or from one to another. Refurbishment projects are very different. Unless extensive exploratory works are carried out, and/or drawings are available from either the original construction date or an earlier refurbishment, it is unlikely that full drawings will be available. Even then checking will need to be done, for often the building is not built as drawn, either because it never was or because it has been subsequently changed or decayed.

There is much to be said for describing the work to be done in words in the form of a *specification*. As the word implies, this means being *specific* about the work, including sufficient detail on such as:

■ what is the work to be done;
■ where is the work and how extensive;
■ when or by when is it to be started or completed;
■ by whom, or by what kind of person, is it to be done;
■ how is it to be done.

Now it might be possible to spell out some or all of the above in great detail. That may or may not be important. It may be vital if, for instance, someone is to carry out work at a distance and is unable to visit the building. Even then a drawing may not convey readily the complexity of an 'uneven' building with its acquired movements and idiosyncrasies. Photographic images are much more readily taken and shared electronically than in earlier days.

I suggest that the repair philosophy or principles be articulated as clearly and succinctly as possible. This could include such as for instance:

- Damaged or decayed work is not to be repaired or replaced solely on the basis of that damage or decay, but because it is injurious to the use of the building.
- Existing material to be reused as much as possible.
- All work to match to existing as far as reasonable practicable, and if not, it should contrast.
- Only appropriately skilled and qualified craftspeople to be used (except as labourers?).
- Wherever rot or rust (or other deleterious material) is identified, all affected material is to be cut out and disposed of responsibly.
- Before and after photographs are to be taken, dated and placed on record in an agreed manner.
- Contemporary standards are to be applied wherever possible without detriment to the building or its architectural quality.
- The client's representative is the ultimate arbiter of what is satisfactory.

One of the problems of course is that there is always scope for inaccuracies, misunderstandings, terminological inexactitudes, of different perceptions of just what was expected. That is a value of standard clauses, definitions of terms, phraseologies, taxonomies and so on; and of working together with colleagues to build up such understandings, experience and a firm foundation for trust. Arguably the smaller the works and the more they are similar but different in nuance, the greater the advantage of 'knowing' intuitively or 'feeling' what is required. Terms such as 'as far as possible' or 'reasonable' or 'as practicable' or 'to match' are always going to be open to alternative interpretation.

The National Building Specification (NBS) was started in 1973 as a common set or catalogue of specification clauses from which to select to describe materials and workmanship for a construction project. NBS builds on British Standards, Building Regulations and codes and guidance from authoritative sources to provide a solid base of standard specifications that have stood the test of time. It saves the time and anxieties associated with working up specifications 'from scratch'. It also enables consistency and continuity of understanding to be built up especially in relation to repetitive work. NBS has developed 'NBS Scheduler' as a package to enable online development of specifications specifically for refurbishment projects (www.thenbs.com). Other products are available.

Summary

This chapter was intended to focus on how buildings may be assisted to provide the required functions more closely for longer. To that end it examined how defects could be eliminated, or at least reduced, by attention to detail at the design and execution stages. This implies a degree of care to be applied, based on an attitude of wanting to do a good job, for personal satisfaction and for the benefit of building users. The following chapter builds on this to consider how work might be organised to best effect.

References

Atkinson, A. (1998) *The role of human error in the management of construction defects.* Proceedings of COBRA '98: RICS Construction and Building Research Conference, Oxford Brookes University. RICS, London, Vol. 1, pp.1–11.

British Standards Institution (1991) BS 4778-2: 1991 *Quality Vocabulary – Part 2 Quality Concepts and Related Definitions.* BSI, Milton Keynes.

Building Research Establishment (1982) *Quality in Traditional Housing, Vol. 1: An Investigation into Faults and Their Avoidance.* HMSO, London.

Burt, M. (1978) *A Survey of Quality and Value in Building.* BRE, Watford.

Cole, R.E. (1999) *Managing Quality Fads: How American Business Learned to Play the Quality Game.* Oxford University Press, New York.

Construction Project Information Committee (2003) *Production Information: a Code of Procedure for the Construction Industry.* RIBA, London.

Coordinating Committee for Project Information (1998) *Common Arrangement of Work Sections* (2nd edn). CCPI, London.

Egan, J. (1998) *Rethinking Construction: Report of the Task Force to the Secretary of State for Environment, Transport and the Regions.* DETR, London.

Flanagan, R. & Jewell, C. (2005) *Whole Life Appraisal.* Blackwell Publishing, Oxford.

Hollis, M. (2006) *Pocket Surveying Buildings* (2nd edn). Royal Institution of Chartered Surveyors, Coventry.

Jaggar, D., Ross, A., Smith, J. & Love, P. (2002) *Building Design Cost Management.* Blackwell Publishing, Oxford.

Kirkham, R. (2007) *Ferry & Brandon's Cost Planning of Buildings* (8th edn). Wiley-Blackwell, Oxford.

Mozer, C. (1984) Total quality control: a route to the Deming Prize. *Quality Progress*, 17, 30–33.

Wood, B.R. (2003) *Building Care.* Blackwell Publishing, Oxford.

Useful websites

www.alway-associates.co.uk/legal-update/article.asp?id=10. Accessed 19 February 2009.

www.rics.org?Practiceareas/Builtenvironment/Constructionmangement/Practical%20Completion%20Defining%20the%20Undefined.html. Accessed 19 February 2009.

www.thenbs.com/products/nbsScheduler/index.asp. Accessed 19 February 2009.

9 Organising maintenance works

The previous chapters were focused on how buildings deteriorate and how they may be assisted to provide the required functions more closely for longer. To that end it examined how defects arose and how they could be eliminated, or at least reduced, by attention to detail at the design and execution stages. This implied a degree of care to be applied, based on wanting to do a good job.

This chapter sets out to examine how work might be organised to best effect. Consideration will be given particularly to the place of statutory control and guidance and the involvement of contractors. The value of supervision and inspection will also be explored. The overall aim is to promote methods that are most likely to achieve the desired end-result of satisfactory buildings in which to live, work and play.

Organisation

As a word, 'organisation' can mean *an* organisation – a firm or enterprise of some kind – or it can mean the process of getting something organised; both uses are appropriate here. Different organisations will organise things differently; every organisation is different. No two local authorities will plan or carry out their building maintenance in the same way; no two housing associations or National Health Service (NHS) trusts. There may be examples where a senior member of staff transfers from one organisation to another and transfers a way of working into the new organisation but the similarity is likely to last only a while before one or other of the organisations rearranges its practices again. There can be a tendency to think that a new manager is not really earning his or her keep unless they are rearranging how things are organised.

The other significant way in which two organisations may have their building maintenance organised in the same way is if they have outsourced the provision of the service to the same company. That company may have developed a way of working that is either particularly appropriate to both organisations, or it may be that it has developed one way that it applies to all its maintenance contracts.

Figures 9.1 and 9.2 are intended to represent the flow of considerations explored in this chapter; they are not meant to be definitive (especially as procurement methods and contract forms are constantly developing and changing) but are indicative of the range and interaction of considerations.

Direct labour

Most public bodies, and some (generally large) private organisations, have directly employed staff of their own, specifically available to undertake building maintenance work. These may

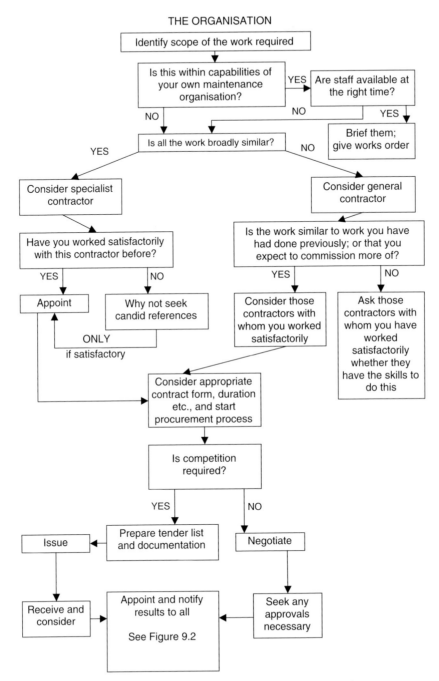

Figure 9.1 Organisation of building maintenance works: in-house or outsourced.

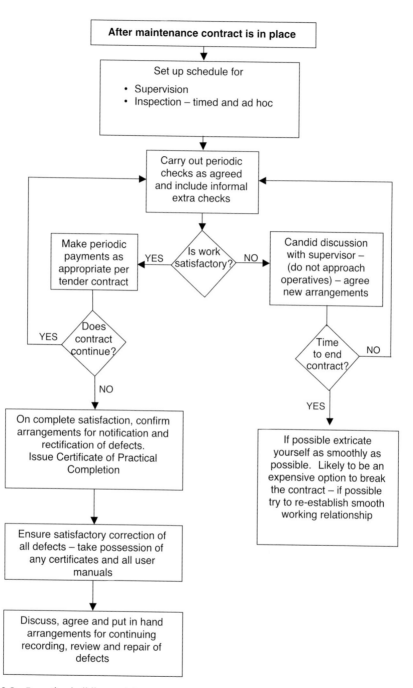

Figure 9.2 Executing building maintenance.

range from general, multi-skilled, or unskilled, handyman to a team of operatives with a range of skills. They may provide a 24/7 service, or may be 9–5 (more likely 8–4) Monday to Friday. These individuals or teams are generally known as direct labour or direct labour (or service) organisations (DLOs or DSOs), because the staff are directly employed as part of the client organisation. Often such workforces have existed as part of those organisations for a long period and may be expected to be resistant to the engagement of private contractors and to outsourcing, whether of part or the whole of the maintenance operation.

Where such arrangements exist, it is important that all members of staff are kept fully occupied or nearly so – overheads are being incurred whether the individual is gainfully occupied or not. Issues related to sickness and holiday cover, and out of hours working will need also to be considered. It may be that a combination of direct labour and contractors will be the most appropriate arrangement.

Specialist contractors

Most organisations are unlikely to carry specialists within their DLO, if they have such an organisation at all – there is not generally enough specialist work to keep any particular specialist operative fully employed. Most direct labour is fairly generalist. Quite a lot of work in existing buildings, especially in buildings listed as being of architectural or historic interest, needs to be carried out by a specialist. Often such people operate as self-employed individuals or in small firms of people of similar or related skills. Less commonly, for instance in an organisation with a large number of old properties such as the National Trust, there may be scope for and appeal in maintaining a sizeable complement of specialist building maintenance staff, together with apprenticeships and appropriate staff and skill development and training programmes.

For any organisation with one or more old buildings it will normally be appropriate, and advantageous when an emergency strikes, to have a relationship established and arrangements in hand with the requisite specialists. A frantic thumbing of Yellow Pages at the time disaster strikes may not deliver the desired result. Good specialists are usually in great demand. They generally ensure a continuous flow of work for themselves by maintaining a full forward order book and a waiting list. When work is plentiful, prices will rise; and they will fall when there is less. No specialist can afford to be out of work for long periods; as self-employed people they are not eligible for unemployment benefit (or job seekers allowance), and they need to build up their own pensions.

What kind of activities may be thought of as specialist? These would be anything that requires a particular skill or experience or qualification; anything that just cannot be entrusted to an unskilled person as the job or its outcome would be dangerous, potentially destructive or otherwise unacceptable, if not done correctly. That is quite a lot of construction-related activity. Indeed, much construction work, including repair and refurbishment, is carried out by specialists engaged either as small firms or self-employed individuals. This is often done on a subcontract basis within the context of a main contract made between the client and a general contractor (who carries the overall responsibility when things go wrong, as they unfortunately often do).

Specialist contractors may be engaged for such work as:

■ lead or copper roofing;
■ slating, tiling, thatching;
■ brickwork, especially when complicated;
■ external rendering;
■ stonework;

- curtain walling and glazing systems;
- plasterwork, especially when decorative, e.g. covings, cornices, mouldings;
- timber panelling;
- marble floor and wall finishes;
- lift and escalator installation and servicing;
- gas, electrical and telecommunications systems;
- underpinning foundations.

General contractors

Most building contracting companies could be described as general contractors. This extends from the most commonly sized enterprise, which is a small business of between one and five employees, to the largest multi-national operation. These companies will offer to undertake any kind of construction work, although many will favour a particular size of contract for which their head office infrastructure is best suited. Most contractors are not specialists as such, although some may have some specialists, for example a stonemasonry or specialist joinery division. Some contractors however may specialise in certain kinds of work, for instance housing or infrastructure or Private Finance Initiative (PFI) projects.

Advantages of a general contractor may include:

- ability to tackle (almost) any task;
- ability for the contractor to find the required personnel (numbers and skills) at the right time by subcontracting as and when required;
- lower price, as the contractor can deploy staff fully and contain overheads;
- greater time-related certainty due to general availability of staff.

Potential downsides include:

- 'Jack of all trades; master of none';
- Overstretch – expectation that they can do everything, and immediately;
- Inconsistency;
- Poor quality of work by non-specialists.

Selecting contractors

The flow chart (Figure 9.1) suggests a step-by-step approach to deciding on a contractor to undertake building maintenance work. Like all models this is fairly mechanistic and something of an over-simplification of the process; decision making is rather more nuanced in practice. Key considerations are quality and reliability:

- Have you worked satisfactorily with this contractor recently?
- Is their work good?
- Did they start and finish on time? If not, why not?
- What are their priorities?
- How would you describe their attitude?
- Have they the capability?
- Can you trust them?

You need to be able to satisfy yourself and others about this decision. Collect and record evidence on which to base your decision, and have someone else – a 'critical friend' or a committee of some kind – to question and support or reject your proposal; there is a lot at stake in this decision. If you are uncertain, consider making a smaller decision – entrust the contractor with a small contract as a pilot project and see how that goes. If the contractor is keen they will do their utmost to impress. If they are not 'up to the job' you will be pleased you only gave them a little job! Of course success on one project is not of itself a guarantee of success on the next, but it will give you evidence, a benchmark, and enhanced confidence that it could and should turn out satisfactorily.

> *Well done, good and faithful servant! You have been faithful with a few things; I will put you in charge of many things.* (Matthew ch. 25, v. 21)

Procurement

This section is about methods for obtaining the maintenance service desired – it is more about the how (with a little about when, where and why) rather than who and what. A range of related issues will be considered:

- prices and pricing;
- schedules of rates;
- tendering;
- contracts;
- service level agreements.

Building Maintenance Information (BMI, 1998) and the Royal Institution of Chartered Surveyors (RICS, 2000) have reviewed and made recommendations on procurement practice for building maintenance works.

Prices and pricing

Nice as it may be to think of money as no object, generally speaking a client has no wish to pay more for an item than was necessary to obtain the required quality of product. *Very* occasionally it may be the case that money is no object – for instance where a replacement item just has to match the original and the client has the ability to fund it – in which case enjoy that rare pleasure. This entails knowing, therefore, what is a reasonable price or the going rate.

A reminder may be appropriate here of the differences between price, cost and value. The price of something is the sum of money charged for a particular item, product or service by its supplier. That price is generally made up of several components:

Price = cost of materials incorporated (+ materials wasted as offcuts etc.) +

cost of labour + proportion of overheads + profit.

You can see immediately that the price is influenced by a number of significant variables. How much material is used and of what quality? How much is wasted, for instance in material that could not be used, or could not be sold because it had perhaps 'gone off' due to poor storage or inadequate manufacture? How much labour was actually applied? Could someone cheaper, made in ascertaining what it is that the customer wants and needs, and ensuring that they get that, consistently.

Issues

The supermarket had reviewed its purchasing policy regarding chilled food cabinets and refrigerated units – significant components in the buying environment. Rather than commissioning manufacture to their own rigorous design standards and specifications it had been decided to buy 'off the peg' on the basis of measured performance in operation – initial and operating costs, reliability, etc.

At the checkout, every purchase that a customer makes is recorded. The barcode conveys the size and type of bread, when it was baked, what shelf it was on, when it was sold and for how much, and – if a credit, debit or loyalty card is used – to whom. That data can be interrogated daily – and it is – to determine patterns of sale and to make adjustments as necessary. Orders for restocking are assembled automatically, and delivered on a 'just-in-time' basis, which is just as well as there is no 'back' area for storage. This information transfer infrastructure was available to convey real-time information also on the performance of the building. Thus, by linking in to a building and energy management system (BEMS), it was possible to monitor when the temperature on a chiller unit was rising toward an unacceptable point.

Findings

The supermarket chain was using the data collected to inform its decisions on maintenance of the buildings and their services. Term contracts were being developed and tested which offered long-term relationships with small local contractors who could offer reliable and speedy response. Typically 3-year contracts were on offer with a 1-year break clause on either side and with opportunity (and expectation) of extension given satisfactory performance. SLAs were in preparation, to be informed by service times that were realistic and capable of consistent delivery. Through trust and open and honest sharing of aspirations and concerns, it was possible to gradually reduce response times and to move from target times for attendance on site to target times for fixing the item.

So, rather than finding extensive and effective PPM programmes, what we found was a very effective system of response maintenance. We dubbed this 'just-in-time maintenance', as it was possible, through use of the monitoring capability of the building management system, to anticipate the need for a maintenance intervention before the item had actually failed in service.

Wider application

These findings seemed to be worthy of wider dissemination and that benefit would be gained by sharing of information on good practice. Other supermarket chains were happy enough for us to visit, and to hear of what their rivals were doing, but not to share what they were doing. In a cut-throat world of low margins there is a lot at stake. Competitive advantage, so hard won, is not to be given away lightly. Maybe naively, as academics we perceive value in the sharing of research findings, and that does seem to work better in the public sector, where the pursuit of best value and the 'beacon council' scheme has been bearing fruit. Perhaps growing interest in energy costs and use of benchmarking clubs may offer possibilities for learning from the innovative approaches of others.

Statutory requirements

Despite a 'sea change' in practices through and since the 1980s, with a massive move from public sector provision through privatisation, there is still a very strong public interest in regulation and control of activities. Reasons for this are many and varied. For some it is important that the individual ought not to be able to gain, perhaps substantially, at the expense of 'the public good'. For others it is a legitimate and proper concern for the interest of the individual – particularly in relation to health and safety for instance. People ought not to be allowed to put themselves, let alone others, at risk of serious injury or death.

Requirements in three main areas are considered here: development control; building control; and matters of health and safety.

Development control: town and country planning

There has been control of development in UK since the 1947 Town and Country Planning Act. There was demand to prevent the swallowing up of the British countryside as there had been by urban, or suburban, estates of houses and factories along the new arterial dual carriageways such as the Great West Road out of London and the Kingston Bypass. Green belts and new towns were promoted and planned. Later there was a keenness to control the redevelopment of run-down areas and the exploitation of 'little old ladies' by rapacious property developers buying their property cheap, obtaining planning approval for a new office block and selling on at greatly increased value. The property boom of the 1950s and 1960s is well described in a book by Oliver Marriott (1967). This boom brought the concept of betterment, whereby the developer made a payment in recognition of that gain in value, and those that were 'losers' might receive compensation. Policies and legislation oscillated with changes in political control in Westminster, and betterment in the guise of Development Land Tax disappeared from the statute book in 1985.

The loss of old town and city centres to redevelopment also prompted the growth of interest, often as a rearguard action in opposition to plans for modern office buildings and shopping precincts, in historic buildings and urban townscapes. Organisations such as the Civic Trust, the Society for the Protection of Ancient Buildings and many preservation trusts grew. The Civic Amenities Act was passed in 1967. Buildings may be listed because they are of architectural or historic interest, either individually or as a group, and conservation areas can be designated. Buildings in these situations are accorded additional protections.

The underlying constant with regard to planning control is that *all* development requires planning permission. Certain kinds of development, however, are deemed to be permitted development. This was defined under the General Development Order and was intended to reduce the load on local planning departments. Thus, for instance, the repair or replacement of rainwater guttering will not normally require planning permission; listed building consent may be required however. A dormer window or a single-storey extension may be acceptable in some circumstances and not in others. The government has recently consulted on granting exemptions for 'householder applications', but seems to be proposing stricter controls of some aspects.

By and large, any proposal to add to a building or to change its use or to alter its outward appearance is likely to need planning permission. To carry out changes that needed permission without having gained that approval is illegal; it is insufficient to have an informal 'understanding' or to be awaiting approval of an application already made. Obtaining planning permission is a time-consuming activity, and requires payment of a fee to the local authority for consideration of your proposals. The key aspect regarding planning is to develop a good working relationship relation to building maintenance contracts, both client and contractor know that there will be a continuing need and therein the possibility of continuing that relationship if both parties so desire. In this case it would be eminently sensible to review how well things went, and arguably negligent not to do so.

It is important to be realistic and candid in feedback without being rude or unhelpful. Although informal feedback has its value, there is potentially greater value in collecting feedback to a consistent format. A combination of open and closed questions, and not too many, tends to work well. Closed questions typically seek responses that are yes/no or look for a rating on a Likert-type scale such as highly satisfied/satisfied/neutral/dissatisfied/highly dissatisfied. If the same questions are asked of different participants in a project or across projects or over time, then triangulation can be done and trends projected.

While this feedback may be useful as a 'wrap up' on a project, it is most useful when applied in thinking about what to do on the next contract – what processes and products to retain and what to change. This turns all that is done into a potential learning opportunity.

Summary

The previous chapters were focused on how buildings deteriorate and how they may be assisted to provide the required functions more closely for longer. They examined how defects arose and how they could be eliminated, or at least reduced, by attention to detail at the design and execution stages.

This chapter set out to examine how work might be organised to best effect, considering the involvement of contractors, the role of statutory controls and guidance and the value of supervision and inspection.

The six chapters that follow will consider defects in more detail. They look at the principal elements of buildings and what may go wrong in and on them. How such defects may be corrected, and how they may be avoided in future, will also be discussed.

References

Baiche, B.M., Walliman, N. & Ogden, R. (2006) Compliance with building regulations in England and Wales. *Structural Survey*, **24**(4), 279–299.

Blanchard, K., Zigarmi, P. & Zigarmi, D. (1986) *Leadership and the One Minute Manager*. Collins, London.

British Standars Institution (1993) *BS 3811: 1993 Glossary of Terms used in Terotechnology*. BSI, Milton Keynes.

Building Maintenance Information (1993) *Measured Term Contracts: Special Report 193*. BMI, Kingston, Surrey.

Building Maintenance Information (1998) *Review of Maintenance Procurement Practice: BMI Special Report 270*. BMI, Kingston, Surrey.

Carillion (2003) *PSA Schedule of Rates for Property management: Composite Items for Maintenance, Repair and Improvement Works*. The Stationery Office, London.

Carillion (2005) *PSA Schedule of Rates for Building Works* (9th edn). The Stationery Office, London.

Clarke, K. (1992) *Measured Term Contracts: An Introduction to their Use for Building Maintenance and Minor Works*. CIOB, Ascot.

Construction Industry Research and Information Association (2003) *Safe Access for Maintenance and Repair*. CIRIA, London.

Egan, J. (1998) *Rethinking Construction: Report of the Construction Task Force.* Department of the Environment, Transport and the Regions, London.

Holmes, R. & Mellor, P. (1985) *Maintenance Coding and Monitoring: Two Case Studies: Technical Information Paper 53.* Chartered Institute of Building, Ascot.

Joint Contracts Tribunal (1989) *Practice Note MTC/1 and Guide.* JCT, London.

Joint Contracts Tribunal (2007) *Measured Term Contract Guide.* Sweet & Maxwell, London.

Latham, M. (1994) *Constructing the Team.* HMSO, London.

Marriott, O. (1967) *The Property Boom.* Hamish Hamilton/Pan Piper, London.

McGregor, D. (1960) *The Human Side of Enterprise.* McGraw-Hill, New York.

Navon, R. & Berkovich, O. (2006) An automated model for materials management and control. *Construction Management and Economics*, **24**(6), 635–646.

National Joint Consultative Committee (1989) *Code of Procedure for Selective Tendering.* JCT, London.

Prior, J.J. & Nowak, F. (2005) *Repair it with Effective Partnering: Guide to Contractual Relationships for Cost-Effective Response Maintenance.* BRE, Watford.

Royal Institution of Chartered Surveyors (2000) *Building Maintenance: Strategy, Planning and Procurement.* RICS, London.

Smyth, H.J. & Wood, B.R. (1995) *Just in time maintenance.* Proceedings of COBRA '95 RICS Construction and Building Research Conference, Edinburgh. RICS, London, Vol. 2, pp. 115–122.

10 Defect recognition and rectification

General

The previous chapter set out to examine how work might be organised to best effect, considering the involvement of contractors, the role of statutory controls and guidance and the value of supervision and inspection. The chapters before that were focused on how buildings deteriorate and how they may be assisted to provide the required functions more closely for longer. The chapters explored how defects arose in general and how they could be eliminated or reduced by attention to detail at the design and construction stages.

The six chapters that follow will consider defects in more detail. They look at the principal elements of buildings and what may go wrong in and on them, how such defects may be corrected and how they may be avoided in future. This chapter applies considerations from previous chapters and sets out general principles and practical procedures related to inspection and recording of defects.

Inspection

My wife reminds me from time to time that God gave me two ears and one mouth, and they should be used in that proportion! He also gave us two eyes with which to see, a nose with which to smell, hands with which to touch and feel and a brain. We are to engage all our senses, and good sense, in observing what is going on in a building and what if anything should be done about it.

In order to see, we need to look; and to hear, listen. These are not just passive tasks; they are active too. We need to engage purposefully. We also need to see what we were not looking for, and to not prejudge by seeing what we expected to see. Inspection can be a very tiring activity, because it requires that level of engagement. It therefore needs to be undertaken soberly, both in the sense of serious and purposeful observation and also without being under the influence of alcohol. Much inspection is carried out in risky places (alongside roads, on roofs and in basements, in unoccupied and potentially dangerous premises) so alertness to hazard is imperative at all times. It is important therefore to be well organised.

Whenever you go out, especially alone, be sure to let someone know where you are going and when you expect to return. Take a mobile phone with you. Remind yourself of the important aspects of undertaking an inspection; these were outlined in Chapter 3. A number of good guides exist (e.g. Bobbett ,1995; Watt & Swallow, 1996; Royal Institution of Chartered Surveyors (RICS), 1997; Hollis, 2006).

Why?

Before visiting site to undertake an inspection, be clear that you know why you are going. This could have an enormous difference to what you do when you get there, and for that matter what you should do in preparation. If for instance you are to inspect a particular property in order to make a RICS Homebuyer's Report then you will need to:

- be familiar with the content, and limitations, of such a report;
- be certain of the correct address; and whether all buildings there are to be included;
- have some early indication of size of the property and other details sufficient to gauge time and personnel involved;
- be sure you have the necessary expertise;
- ensure access is available and that you are expected;
- ascertain what kind of testing might be required, e.g. of drains, services;
- be properly equipped and supported;
- be sure you have the client's instruction confirmed in writing.

With effective briefing and preparation it should only be necessary to visit the property once in order to do the job required. This is important; it is embarrassing and inconvenient to have to return to undertake work which could and should have been done first time. Treat each project as if the building were at the other end of the country, or the other side of the world, which it could be sometime! Practice and cultivate good habits.

How?

Being methodical and thorough is crucial; it could be very expensive for your organisation if you miss something you should, and could be expected to, have seen and reported upon. Any item which the client has particular concern about should feature prominently in the inspection, together with all areas which are known (often by painful experience) to be particularly prone to problems. Chimneys, flat or complex roofs, parapets, cellars, basements and areas below ground are usually particularly problematic and demand to have time spent on their inspection. Thus, if you have reason to suspect that the building may have parapet walls, you must expect to inspect the roof areas behind, so you will need appropriate equipment and access. It may be that you will need to recommend further inspection beyond your initial inspection.

In order to make recommendations for remedial actions on defects it is important to diagnose and deal with the cause or causes, not the symptoms. Thus, for instance, when inspecting the property with parapet walls one must expect the possibility of leaks from the parapet gutter. The inspection therefore – not only of the parapet wall and the gutter behind it but also of the roof space and the rooms beneath – needs to look for evidence of leaks. Similarly, observation of staining and dampness in a room demands 'following the trail' to consider possible causes. Whilst it is quite possible that the cause of such dampness is water leakage from the gutter, there will be other possibilities to consider. Possibilities could include for example: a leak from a cracked or blocked rainwater hopper or down-pipe; a crack in the external wall; poor detail or construction at a window lintel; absorptive materials; leak from an internal water pipe or condensation on it or on or in the wall. Any or all or none of these may be the cause, or a contributory cause together with other deficiencies. This situation is discussed in more detail in Chapters 12 and 13. Suffice it to say here, generalising, that to identify one possible cause is not the same as having identified *the* cause; nor is having identified a cause the same as having identified *all* the contributory causes.

What is required at the time of the inspection is to observe enough as to be able to decide there and then, if possible, what subsequent action should be recommended. The making of sketches and taking of photographs will assist recording, reporting and subsequent discussion and deliberation. It may be that the decision taken on site is that the matter should be considered at greater length in the office later. This may need to be informed by the collection of further data, perhaps involving 'opening up' of parts of the building for more detailed inspection and/or tests.

The inspection should be carried out with all senses and brain engaged. Thus when observing the stained and peeling wallpaper, you also look and smell for mould, take and record moisture meter readings, feel the surfaces, think 'where else might this occur?' Look for other instances, consider what might be causes and follow the trail. Also think about what could be other effects, where and how they might be manifested and follow them up too.

Anything that looks or feels unusual, unexpected, out of place, not flat or not level demands inspection. It may be that the apparent defect has been there from the original construction, or is of long-standing. It may be benign; it might not be. Alterations, cracks, deflections, stains and previous repairs are all indicators of areas worthy of particular attention in the inspection.

When?

Buildings should always be inspected:

▪ before purchase;
▪ before and after the completion of building-related work;
▪ before issue of a certificate of practical completion;
▪ before issue of a final certificate;
▪ periodically; some parts annually; some parts more frequently (as previously discussed).

In some climates, the time of year will be significant; and for some building uses the time of day and week will also be important.

Although it will have some inconveniences, it is worth trying to inspect a building during or shortly after rainfall. This will help to demonstrate shortcomings in the rainwater collection and disposal systems. This is particularly important as not only have these been consistently common and damaging sources of water ingress, but anecdotally at least there does seem to be increasing incidence of longer and heavier 'thunder bursts'. Rainwater systems are going to need to perform their role well into the future.

There is also something to be said for carrying out inspections while buildings are occupied. Although it may be more difficult and take longer to move around the building and have furniture and finishes such as carpets that cover up some defects, the ability to ask occupants about problems is helpful. It is good to be able to combine both – perhaps an inspection that commences while the building is occupied and continues when workers have gone home or residents have gone to work; or vice versa.

Be careful, though, about working in buildings or parts of buildings that are in use but temporarily unoccupied; you could be accused of theft of some item, perhaps some days later. Be careful too about working unsupervised where old or young people may be present; again you could be accused, however unfounded, with something inappropriate. This could be very damaging to your career. It is a sad indictment of society to have to think about such possibilities, but one must.

Where?

The principal location for defect recognition is of course the building in question, but the laboratory, library and surveyor's office may also have roles. The surveyor must be thorough: every surface must be examined. Technically, this is what is meant by a *superficial* inspection – that is to say a *surface* inspection, not a shallow, cursory or trivial one. Each surface should be inspected, top to bottom, inside and out. This must be done methodically so that no surface is missed inadvertently. Junctions of materials, and anywhere complex, are worthy of special attention; matters are more likely to go wrong at these points. And of course you must 'follow the trail' of anything suspicious – anywhere that looks (or you sense or think) is in any way unusual.

Some materials may lend themselves to testing in the laboratory. Asbestos (or materials suspected of being asbestos), concrete, mortars and timber may warrant laboratory testing of samples. The taking of these samples and their testing should only be undertaken by specialists. Asbestos must only be handled by licensed removers.

Some matters will require further thought at the surveyor's office or in the library. It may be that drawings or specifications thought to relate to the building in question may throw some light on understanding what is going on in that part of the building. Caution is required however; often buildings are not constructed as designed and the building may also have had modifications carried out subsequently. Old construction text books may also illustrate constructional forms and details that were common in earlier days.

Who should inspect?

To an extent, anyone who is methodical, observant and questioning could carry out an adequate inspection, even if they were unqualified or inexperienced. Is that sensible however when the health, well-being and safety of building users is at stake? No qualification is needed for a person to call him/herself a surveyor. A chartered surveyor, however, will have studied at least 3 years for an honours degree accredited by the RICS and passed an assessment of professional competence. An architect will have studied at university for 5 years and had 2 years of approved professional practice before being allowed the use of the title architect. A local authority building inspector may not be qualified at all, although it is likely that he/she will have a building surveying or environmental health qualification.

I would suggest that no building inspection should be carried out by other than a chartered surveyor or architect, and preferably one with experience of the kind and size of building in question and the locale.

What is to be done?

The five chapters following this one explore issues in relation to the various elements of the buildings, such as walls, windows and doors, roofs and floors, building services and external works. They are intended as overviews to the kinds of problems that may be manifested, and what may lie behind them. They follow the pattern:

symptom > possible causes > rectification > avoidance

developed through my work at the then National Building Agency with colleagues at the Building Research Establishment (BRE). My lifetime interest in defects and their elimination stems from my involvement with the Quality in Traditional Housing project (Bonshor & Harrison, 1982).

There are limitations on how much any one book can cover in detail, including this one, so it is important to 'signpost' where one should go for further information and guidance. I have no hesitation in recommending the BRE Building Elements Series (Harrison, 1996; Harrison & de Vekey, 1998; Harrison & Trotman, 2000, 2002; Pye & Harrison, 2003). These allocate more or less a whole book to each of the elements to which I have been able to allocate one of next five chapters of this book. While I concentrate on the overview from the point of view of a maintenance or facilities manager, they give much more detail. I make extensive reference to BRE publications of various kinds including Defect Action Sheets (now withdrawn but containing useful background and information), Digests, Good Building Guides, Good Repair Guides, Information Papers and a range of reports together with relevant British Standards and other authoritative guidance. Remember that knowledge and practice are constantly being added to and updated, so the currency of all guidance must be checked before deciding upon remedial actions. Subscription services such as the online database Construction Information Service (formerly Barbour Index) are invaluable.

Recognition

How does one recognise a defect? It may be something that is not as others of similar detail are – for example an instance of an otherwise repetitious detail that has some part missing or dislodged or broken in some way. It may not be causing a leak or a structural problem but it is a defect still, worthy of recording. On the other hand there may be a defect where all is apparently in place as it should be, but out of sight it is defective in material or fixing or of inadequate thickness or overlap, perhaps even leaking already unnoticed into some void in the structure. How might that be recognised?

This is where experience is very useful. Over years of practice one develops both knowledge of where defects have been found but also a 'sixth sense' of where a building may be susceptible to defects. More time should be allocated to inspection of those areas most likely to have defects, and those where defects are most likely to be able to have serious effects. Also, the more one looks, the more one tends to find, so appropriate 'quality' time must be allocated. What does one actually *see*? It is said that 'seeing is believing'; I suggest that also 'believing is seeing'. Once sensitised to the possibility of a defect, one is more likely to see it. It may well be worthwhile to have small samples tested in order to ascertain whether the perceived defect is in fact so; both client and surveyor will be relieved to receive a supportive 'all clear'. The seminal detailed work on defects by Eldridge (1976) (updated by Carillion, 2001) should be recognised and the works of Hinks and Cook (1997) and Marshall *et al.* (1998) are informative.

Surfaces on which water falls, and particularly where it collects, are especially susceptible to damaging water penetration. Areas in contact with the ground are likely to suffer from ground movement. Elements exposed to the sun will experience diurnal variations in temperature. All these situations suggest a need of inspection of vulnerable details. Any complex or unusual construction must be considered a potential source of a defect and inspected accordingly. Any disruption of a surface, or unevenness, or difference from what is around or from what was expected generally must be considered with suspicion. If it looks odd, it possibly is!

Recording

Date, time, weather conditions and surveyor must always be recorded.

Some surveyors like to record their observations directly on site into a hand-held voice recorder; this is a good practice. Alternatively, or additionally, notes should be written, either by hand on to a pad or a previously prepared pro forma or using an electronic equivalent, perhaps a palmtop computer. These should be a definitive capturing of what the surveyor saw and considered at the time. It is no good trying to fill in missing observations later on in another place. If vital information is missing a return visit (with consequential extra time, cost and inconvenience) will be necessary.

These notes should be checked for completeness and clarity before leaving the roof, the room, the roof space, the building, the site. Sometimes a kind of shorthand may be used; if it is, be sure that it is comprehensible by anyone who may be called upon to transcribe the notes, so that 'translation' errors do not creep in.

Photographic records are particularly useful, relatively cheap and effective. The locations of each must be plotted on to a plan or a sketch made to show the locations, and each numbered for clarity. Again it will be infuriating, and difficult, to try to 'pin down' locations of photographs a week or so after the event. It is particularly useful to have photographs which show the extent of defects such as cracking and deflection, leaking and ponding. Often a combination of locational and detail photographs will 'tell the story' most effectively. Even with the quality of modern equipment it is worth trying to compose the shot with light and shade so that the three-dimensional arrangement and relationships are clear. Sometimes a coin or other indicator of scale or size is worth including in the picture too.

Reporting

A report on defects to the client will not always include every note and detail recorded at the time of investigation. It must however tell the complete 'story'. While, for instance, notes and photographs will have been recorded as the surveyor transited through the building, it may make more sense to present the report elementally or thematically. Thus all matters related to roofs may be reported together or all those related to water or w.c. provision. The important matter is to convey what needs to be conveyed in a well organised, logical and authoritative manner that will be effective in eliciting the required response, for instance an allocation of (additional) funds to deal with the important matters.

A report is for a purpose, and that must be kept in mind when writing and organising it. Presentation is important. It is said that 'a picture is worth a thousand words' and it is true. Some people understand pictures much better than words, so include photographs – and caption them so that the reader knows what you intend the picture to convey – don't leave it for the reader to draw the wrong conclusion! Others will relate to graphs and charts or to statistical analysis; for yet others the 'bottom line' will be all that matters, so be prepared. It can be useful to show how technology it may be that this inspection and approval can be carried out remotely and the pictures also downloaded remotely and stored for record purposes.

Follow-up

Inspect work when it is nearing completion. Satisfy yourself as to whether any further work, however unexpected, may be required or is desirable. Take photographs and make notes for the record as appropriate. Agree contact details and a date with building users and contractors when the works will be re-inspected, not only for latent defects, but also to review how well the specifications and associated processes have worked out in practice. This will help the making of future decisions.

Checklist: defect recognition and rectification

Figure 10.1 shows the process diagrammatically; and some of the boxes are developed further below as an indication of some of the kinds of ways that defects may be apparent and considerations that may be made.

A defect is apparent because:

- Something looks wrong; for instance:
  - □ something is misaligned;
  - □ or misshapen;
  - □ or broken;
  - □ or discoloured;
  - □ out of level or off-vertical.
- Something smells strange, such as:
  - □ gas leaking;
  - □ rotting.
- Someone heard something:
  - □ rubbing;
  - □ hissing;
  - □ bubbling;
  - □ gurgling;
  - □ cracking.
- Something felt:
  - □ rough or pitted;
  - □ uneven;
  - □ out of level.

Summary

This chapter has established general principles and practical procedures related to inspection and recording of defects. Observations have also been made on the need for decisions to be informed by experience and for these to be followed through into the preparation of proposals and recommendations for defect rectification. In the following chapters, these processes are applied in turn to the main elements of buildings.

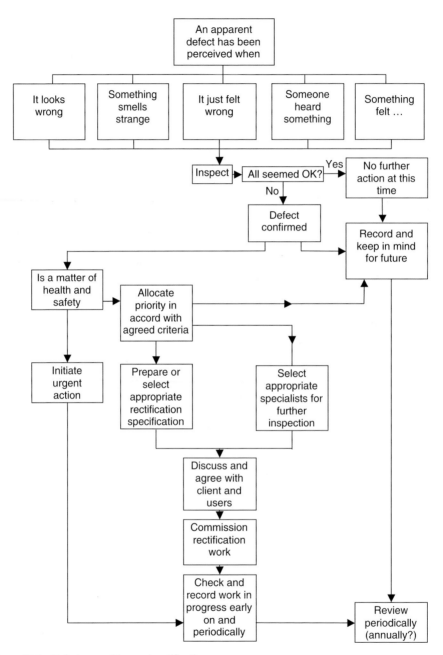

Figure 10.1 Defect recognition and rectification.

References

Bobbett, I. (1995) *Condition Assessment Surveys: BMI Special Report*. Building Maintenance Information, Kingston, Surrey.

Bonshor, R. & Harrison, H. (1982) *Building Research Establishment Report: Quality in Traditional Housing*. HMSO, London.

Carillion (2001) *Defects in Buildings: Symptoms, Investigation, Diagnosis and Cure* (3rd edn). The Stationery Office, London.

Eldridge, H.J. (1976) *Common Defects in Buildings*. Department of the Environment/Property Services Agency, London.

Everett, A. (1994) *Mitchell's Materials*. Longman, Harlow.

Harrison, H. (1996) *Roofs and Roofing: Performance, Diagnosis, Maintenance, Repair and the Avoidance of Defects*. BRE Press, Watford.

Harrison, H. & de Vekey, R.C. (1998) *Walls, Windows and Doors: Performance, Diagnosis, Maintenance, Repair and the Avoidance of Defects*. BRE Press, Watford.

Harrison, H. & Trotman, P.M. (2000) *Building Services: Performance, Diagnosis, Maintenance, Repair and the Avoidance of Defects*. BRE Press, Watford.

Harrison, H. & Trotman, P.M. (2002) *Foundation, Basements and External Works: Performance, Diagnosis, Maintenance, Repair and the Avoidance of Defects*. BRE Press, Watford.

Hinks, J. & Cook, G. (1997) *Technology of Building Defects*. E. & F.N. Spon, London.

Hollis, M. (2006) *Pocket Surveying Buildings* (2nd edn). RICS Books, Coventry.

Housing Association Property Mutual (1995) *Component Life Manual*. E. & F.N. Spon, London.

Marshall, D., Worthing, D. & Heath, R. (1998) *Understanding Housing Defects*. Estates Gazette, London.

Pye, P.W. & Harrison, H. (2003) *Floors and Flooring: Performance, Diagnosis, Maintenance, Repair and the Avoidance of Defects* (2nd edn). BRE Press, Watford.

Royal Institution of Chartered Surveyors (1997) *Stock Condition Surveys: A Guidance Note*. RICS Books, London.

Son, L.H. & Yuen, G. (1993) *Building Maintenance Technology*. Macmillan, Basingstoke.

Watt, D. & Swallow, P. (1996) *Surveying Historic Buildings*. Donhead Publishing, London.

11 Defect recognition and rectification
Foundations, basements and external works

The preceding chapter established general principles and practical procedures related to inspection and recording of defects. Observations were also made on the need for decisions to be informed by experience and for these to be followed through into the preparation of proposals and recommendations for defect rectification.

This chapter is the first of five that apply those survey and recording techniques to elements of buildings. I have chosen to start with a consideration of the lowest part of the exterior of the building with a view to moving methodically up the exterior then proceeding internally from top to bottom.

Outline: general arrangement

Each chapter follows a methodical progress: from the general to the specific, from top to bottom (or vice versa), inside-out. For each element of the building, generic defects are identified and discussed. No single book such as this can cover all possible defects that may be found; nor can it provide solutions to every eventuality – each is different in its context. Guidance is given to further sources of advice and guidance, pointers to ways toward rectification.

Individual defects are discussed in a consistent format:

Symptoms > possible causes > rectification > avoidance.

Symptoms

This is what the surveyor or person reporting the defect sees, notices or senses – it is often no more than a surface observation. It is what is manifested – no more, no less; that which can be reported. Just like you might perceive as a medical condition that which you might report to the doctor; something that isn't as it was, or not as you think it should be, or doesn't feel right. The 'job' of the patient is to report to the doctor as fully and accurately as possible just what it is that is seen or felt, to help the doctor reach a diagnosis of what may be going on underneath and possible causes so that a cure can be proposed.

A symptom is not a problem; it is a sign, a pointer to a possible problem. Symptoms might include for instance:

- cracks or crazing;
- breakages;

The most common manifestations of foundation problems are cracks in walls, floors out of level and/or walls not vertical or with brick courses seen to be sagging.

Cracks

Cracks may be broadly straight, vertical, horizontal or diagonal; they may be of fairly consistent width or tapering; they may be increasing in length and/or width over time; they may increase and decrease cyclically – either by day and night or seasonally. They may appear to be shallow – only affecting a surface – or may be deeper; they may be singular occurrences or may display some kind of regular pattern. Observing these features over a period of time may be significant in determining the most likely cause of any particular crack. 'Tell-tales' may be fixed to the building across the crack in order to observe and measure its width and thereby the amount of movement over time. Two BRE Digests (1989b, 1989c) advise on measuring and monitoring movement. Records, including photographic evidence, should be kept.

Cracks are caused by movements that the building cannot accommodate – something has to give. In foundations these are normally caused by movements in the ground, which may in turn be caused by events of weather or climate or by actions of man. The first BRE Good Repair Guide (BRE, 1996a) provides valuable guidance related to cracks caused by foundation movement.

Ground conditions

There is not room here for an in-depth study of soil mechanics; there are many books on the subject, and when detailed consideration is required it will almost certainly make sense to engage the services of a geotechnical surveyor or structural engineer. By engaging a specialist one gains access to a whole career and catalogue of experience, increasing the likelihood of the problem or one quite like it having been encountered and dealt with successfully before, and thereby reducing risk (Jones, 2004).

In brief, soils contain particles of differing sizes and distributions and water in varying amounts. In wetter weather the soil may absorb more moisture and in hotter weather it may dry out – this is particularly noticeable in clay and sandy soils. By and large, soils are found in fairly consistent strata across a site, but not always. Foundations should go down to a stratum that will bear the load of the building, transferring the load safely to the ground beneath. If foundations are seen to be failing, by observing deflection in the building above, clearly some error was made, whether in the assessment of load or of load-bearing capability, or some other change has intervened, such as subsidence from mining or leaking drains or tree roots.

Clay soils

Clay is a cohesive soil; that is to say it requires a certain amount of water within it to maintain its integrity. When it dries out it cracks up. Thus buildings founded on clay soils tend to manifest foundation-related problems around their edges and especially in hot summers. The ground under the bulk of the building tends to be sheltered from the changes in climate or weather and to stay reasonably constant in moisture content. Around the edge, however, that which is exposed to the weather gets alternately wet and dry with the changing weather. The loss of moisture by drying tends to result in cracking. In particularly hot summers this cracking can be substantial and result in turn in cracking of the building as the ground shrinks away and fails to support the foundation.

Clay soils are prevalent in the south and east of England, and these areas of the country are more susceptible to foundation failures than other regions. There were serious and widespread foundation-related problems across the southeast in the exceptionally hot summers of 1975 and 1976, giving rise to substantial claims against builders of new housing. It is not possible, all things considered, to avoid building on clay soils in the southeast of England – land is in too short a supply! A geotechnical or structural engineer should be engaged to assess foundation requirements in problematic areas.

Tips and fill

As with clay, it is difficult to avoid building on tips, however desirable that may be. The so-called developed countries have been generating waste since the start of the Industrial Revolution. There was waste before then too, of course, but it was largely organic and would degrade quickly under the effect of weather over time, and it was not in 'industrial' quantities. With population growth and urbanisation there has been growth in rubbish to be disposed of. Typically this was collected and taken to waste disposal sites or rubbish tips. These tips would include both residential and industrial waste, including all manner of organic, toxic and non-degradable material. All urban areas would have such tips on their peripheries and, as towns grew, they grew closer. Increasing land shortages have resulted in many of these tips being built upon. Sites should be investigated thoroughly (Charles & Watts, 2001).

Manifestations of problems related to building on such sites include:

- subsidence seen in walls;
- floors out of level;
- dips in ground, including roads, paths and parking areas;
- dislocated underground drainage pipes;
- poor and dying vegetation.

Decomposing material gives off methane, a greenhouse gas, which is also poisonous and potentially explosive. It is therefore dangerous if it should leak into and build up in buildings or in ducts or under-floor voids. Where the presence of such is suspected or may be anticipated, for instance where it is known that the building is constructed on a former tip, ventilation should be provided. This should be assisted by such as the stack effect or fan-assisted to ensure that the gas is dissipated and does not build up.

Mining

Some areas are affected by underground mining. There may sometimes be effects from the tunnelling processes, especially related to vibration caused by the digging machinery. More often there are problems of subsidence related to previous mining activities. This can manifest itself in the form of a 'wave' passing through an area, causing buildings to settle as the mine tunnels beneath collapse in on themselves over time. This happens because valuable pit props are removed when the mine is 'worked out' and there is no continuing support to keep up the roof of the mine tunnel. The former National Coal Board, formed when the UK coal mining companies were nationalised in 1946, kept records of mine workings. These are now held by the National Archives (www.nationalarchives.gov.uk) at the Mining Records Office in Mansfield, Nottinghamshire (www.coal.gov.uk/services/miningrecords).

Care needs to be exercised in the use of such records as they may not be entirely accurate or complete; they may not, for instance, include workings related to other minerals such as copper or lead. Geological maps may be of value in indicating the kinds of rock and strata that may be 'suspect' in relation to ground and subterranean conditions.

Trees

On the whole people rather like trees; they have a 'softening' effect on a locality, give shade from sun and shelter from rain and winds, and delight as deciduous trees' leaves change colour and fall in autumn and come again in fresh greens in the spring. But from a buildings and maintenance point of view trees can be a nuisance. They can give problems in terms of fallen leaves blocking gutters and of people slipping on wet fallen leaves on paths (see later). In relation to foundations, it is tree roots that are problematic (BRE, 1996b).

A 'rule of thumb' is that trees should be the same distance from a building as their mature height. However, many trees and buildings happily co-exist more closely than that. For some particularly 'thirsty' varieties, such as oak, elm, eucalyptus/ blue gum, willow, poplar and cypresses, the distance should be greater, perhaps 25% more (National House Building Council (NHBC), 2003). Trees ought not to be cut down just because they are closer than such a rule of thumb, but it may help as a guide to a possible contributory factor in a foundation problem. Indeed, cutting trees down or even pruning them can give rise to problems rather than solve them (O'Callaghan & Kelly, 2005). When trees are removed the movements of water through the ground are disrupted and changed, and there can be heave when water accumulates in some soils such as clay.

Tree roots can also invade drains, entering through cracks and joints, causing blockages and interruptions to flow. Water and effluent discharged from broken drains may contribute to undermining of foundations by washing out surrounding material that should be supporting them.

Many trees are the subject of tree preservation orders and cannot be cut back or removed without permission of the local authority.

Radon

Radon is a naturally occurring radioactive gas; it is given off by uranium-bearing rocks such as granites, limestone and carbonaceous shales. It has been identified as dangerous in that it can build up in buildings in affected areas and cause sickness to inhabitants of houses in those areas. A map provided by the Health Protection Agency (www.hpa.org.uk/radiation) shows the southwest of England, parts of Somerset, Wales, the Cotswolds, Derbyshire, the Pennines and North Yorkshire to have particularly high concentrations.

Measures to take in properties in affected areas include the sealing up of gaps in the structure, especially floors, to prevent entry of the gas, and the provision of ventilation to avert build up of gas in rooms and in voids.

Underpinning

The traditional response to foundation failure is underpinning. There are specialist firms that offer this service. The operation normally takes the form of digging out in phases and in short sections underneath the existing footings and filling in with concrete, reinforced as required by the calculations. Another process involves the provision of raking short piles from around and/or within the building to pass under the existing footings to carry the loads deeper and distribute them more widely. Compact equipment is available to gain access to constricted sites, for instance to reach the rear of buildings between trees and/or buildings or via narrow gateways, or to gain access into the interior of buildings via external door openings and with limited headroom. Clearly such work requires buildings to be vacated for the carrying out of the work.

Underpinning is designed to retain the building in the position it has reached when the work is carried out – that is, to stop it moving further. It is not intended to restore the deflected parts to their original or originally intended position. Picture the Leaning Tower of Pisa – it was still incomplete when it subsided and the latter stages incorporated a little correction in their alignment. The tower has recently been stabilised so that it should not lean any further than it now does. There are proprietary products and systems available for filling voids beneath buildings to assist in stabilising buildings in that way; they should be carefully checked as to what guarantees or warranties are offered. References should be obtained and visits should be made to buildings where the system on offer has been successfully applied. Check if an Agrément Certificate has been issued; that it is in date and applicable to the situation you have; and that any installer is appropriately qualified and experienced.

Avoidance of problems in future

The most important aspect is 'site investigation'. The BRE has produced excellent material on this for many years; testimony to its continuing need and importance (BRE, 1987a, 1987b, 1989a, 1996c, 2002, 2003; Charles & Watts, 2001). Sites ought not to be bought or buildings built without a thorough site investigation. This would avoid inadequate and failing foundations, saving many thousands of pounds of rectification works and the associated disruption. Further steps would be to avoid planting trees (especially varieties that will grow to great size) close to buildings, or in the vicinity of drain runs.

All drainage, including roof drainage, should be so organised as not to concentrate water discharges around vulnerable parts of the building; any blockage or settlement is likely to give rise to water accumulating and possibly washing away soil from around and under foundations. Ponding of water adjacent to external walls is also likely to give rise to dampness, moss and mould growth, externally and internally.

Once problems have been identified in buildings already in the portfolio they should be monitored regularly to ascertain whether they are increasing, or returning after having been dealt with. This will enable possible further problems to be minimised.

Basements and cellars

The most common problems here relate to dampness. Symptoms include visible signs such as moulds and stains, possibly running water, cracks and dislocations in walls, drains and pipes, rot in timbers, sometimes accompanied by other indicators such as a musty smell.

Clearly matters will need to be addressed more fully and urgently if the basement area is inhabited or in regular use of some kind. It may be a basement flat that is affected in some part, whether large or small; this will need to be rectified completely and quickly – occupiers may need to move out until the necessary work is completed. Even if the area affected is only in use for storage, dampness may be destructive. It would be a great shame to lose antique heirlooms or archived material through unsuitable environmental conditions caused by dampness. Matters may not be of such concern if the space is in use as a coal cellar, but if it is used as a wine cellar it would be a great shame to have labels fall off due to dampness or become unreadable due to mould growth.

Some possible contributory causes may be high water table, underground streams and water courses, cracked pipes and drains, tree roots and lack of ventilation.

Water table

Water is held in and passes through the ground in differing ways in different soils and at different times. This book is not intended to address detailed matters of hydrology. Sometimes ground

Figure 11.6 A cobbled street – beautiful but expensive to repair and reinstate.

The components themselves may spall under the effect of frost expansion or of erosion by foot traffic. They may become uneven and out of level as underlying ground compacts under use or subsides.

Handrails and guarding are further elements that may be deficient or fail. Fixings are particularly vulnerable and should be checked periodically to ensure they are secure. They may be fixed to retaining structures which may themselves be in poor condition.

Steps should, wherever practicable, be lit for use after dark.

Especially where steps are in poor condition, consideration should be given to their replacement by ramps; these are more acceptable as more universal access is expected for people with restricted mobility, people in wheelchairs or on crutches, people with prams, and so on. Slopes are still difficult, and tiring, to negotiate, so they should in any case be minimised and have level areas at intervals. These will enable people to pause and take a breath; seats should be considered for these areas.

Figure 11.7 illustrates that modifications to external works may be extensive; here the entire stone paved area is being rebuilt at altered levels to remove the steps that were a hindrance to access for disabled people. Figure 11.8 shows the importance of correct selection and specification of materials.

Figure 11.7 Adjustments to paving levels to improve access for disabled people.

Parking

Parking of cars and other vehicles can provide management problems which may impinge upon maintenance staff and activities. If parking is uncontrolled, then from time to time unlicensed and/or stolen or unroadworthy vehicles will be left there. Responsibility for dealing with the matter (and perhaps disposing of the offending vehicles) will depend upon the nature of the area. If it is an area of adopted highway, then the highway authority and/or the police may take the matter forward themselves. If however, the area is a private one then the land-owner will have to make arrangements through the police to check on vehicle ownership and responsibility. Organisations that carry out wheel-clamping and vehicle removal need to be sure of their legal status and standing. Vehicles may also be damaged.

If parking is controlled in some way, whether by physical barriers, payment machines, marked or designated parking bays or yellow lines, then some kind of enforcement regime is required. People who are paying for parking may also feel entitled to a higher standard of provision, including its maintenance, than otherwise.

Areas designed for light traffic may, over time, be subjected to higher loads, which will take their toll. There will also be occasional need to dig up areas that have so suffered, or had problems with new or existing service runs beneath, resulting in a patchwork of reinstatements. Periodically it may be necessary to address the resultant visual degradation and restore a unity of surface appearance.

Figure 11.8 Selection and specification of materials and details are important.

Where parking is provided within purpose-built structures (multi-storey car parks) or in basement levels of other buildings, there will be periodic damage to be dealt with. Hopefully this will not be structural, but it may sometimes be necessary to seek the advice of a structural engineer and/or a concrete repair specialist.

Soft landscaping

Trees, shrubs, grass and flowers need to be maintained, to varying degrees in different places and at different times with differing degrees of immediacy. Typical examples of such work may include:

- tree, shrub and rosebush pruning;
- tree limbs cracked or fallen;
- leaf sweeping (and gutter clearance);
- grass cutting/lawn mowing;
- creating of flower beds, planting, renewing annual bedding;
- ponds, water courses, ornamental water features (fountains etc.);
- maintenance of wildlife habitats.

Some work requires specialist, qualified or licensed labour (for instance arboriculturalists, tree surgeons) and/or specialised equipment and trained operatives. Much of the work is relatively unskilled and therefore employs relatively low-paid staff, but, because of its regularly repetitive nature, is an extensive and therefore expensive component of an annual maintenance budget. It is also important to 'make hay while the sun shines' – some of the work just has to be done in fine weather!

Sometimes, poor choices of vegetation or inadequate design support a rethink. Some plants do not thrive in some soils, or in too much sun, or shade; they should be supplanted, and removed perhaps to another location. Small areas of grass should be avoided – they are difficult and a nuisance to mow. Some grassed areas get waterlogged or walked over too much; some would be better paved over.

Common areas generally

It is a generalisation of course, but on the whole areas not clearly owned by someone are often unloved, abused and neglected. Common areas can include:

- stairs and lifts in blocks of flats;
- refuse disposal areas;
- shared garden areas;
- shared clothes-drying facilities;
- garage courts;
- play areas.

Such areas may be characterised by such manifestations of lack of love and care as:

- broken glass;
- litter;
- graffiti;
- vandalism;
- rubbish (whether left in bags or strewn around by vandals or vermin);
- hypodermic syringes;
- a general lack of maintenance attention.

Approaches to dealing with such problems could include:

- (more) frequent patrols by street wardens;
- more street cleaning;
- review of duties: 'whose job is it to . . . ?';
- more, or better trained, or better remunerated, staff;
- allocation of higher priority to works to common areas;
- better lighting;
- closed circuit television (CCTV) observation;
- (better) education of local people;
- redesign to 'design out' problem areas;
- privatisation of currently common areas.

Figure 12.1 A crack, wider at the top, indicating rotation of foundation to the extension.

wider at one end than at the other and result from differential settlement or localised subsidence (Figure 12.5).

Differential settlement occurs because ground is variable, with different load-bearing capacities in different parts, and because loadings in different parts of a building may also differ. Where a section of wall is falling away at or up to around 1 m above ground level it is quite likely attributable to settlement in the ground beneath. This may appear in the wall as:

▦ a dip in horizontal bed joints of brickwork where the brickwork is holding together well;
▦ or a dropped section of wall with stepped cracks following brick joints (where the mortar is weaker);
▦ or as a vertical crack below the corner of a window (a point of weakness in the wall), with the crack wider at the bottom.

At damp proof course (DPC) level, if there is one, there may be expansion of brickwork manifested in horizontal movement of the building on top of the DPC, maybe by as much as 20–30 mm. This happens when the brickwork is in longer lengths than would be recommended today (see Brick Development Association, 1986; BRE, 1985) without the provision of adequate movement joints. Some varieties of brick, e.g. calcium–silicate or sand–lime, are more prone than others to this. Figure 12.6 shows an example of oversailing, though it may be that this is an intended design feature.

Figure 12.2   Cracking at the junction of boundary/retaining wall with main structure.

Figure 12.3   Diagonal cracking: possible foundation settlement or localised subsidence.

Figure 12.4 Vertical crack probably due to brickwork movement and rotation of the short return wall.

In some buildings, particularly in former coal-mining areas, horizontal cracks may be present at regular vertical intervals up the building, normally approximately 400–500 mm. These are typical manifestations of wall-tie failure, the spacing being that of the ties. Causes could be rusting of ungalvanised (or poorly galvanised, or damaged) metal wall ties or interaction of the metal with, for instance, black ash mortar, a by-product of coal mining activities. The problem is more prevalent and more apparent in industrial areas and on walls exposed to more rain, especially driving rain in windy areas. Gable ends are more vulnerable, especially where gable triangles are untied to roof structure. Guidance on recognition and rectification has been issued by the BRE (1983d, 1996).

Cracks may be present above doors and windows, for instance:

▦ dips in horizontal bed joints of brickwork due to absence or inadequacy of lintels;
▦ cracks in relation to fallen or otherwise disturbed arches or lintels;
▦ stepped cracks on the diagonal away from corners of doors or windows;
▦ vertical cracks between the corners of doors and/or windows on adjacent levels in the building.

Additionally where windows or doors are at or near roof level there may be further issues, such as:

Figure 12.5 A short vertical crack perhaps related to localised subsidence.

- movements in timber or steel roof structures such as trusses or rafters deflecting or expanding and pushing out walls;
- movements in concrete roofs or poor detailing at junctions with external walls;
- small pieces of wall above windows or doors, relatively unrestrained and more vulnerable to movement, including lack of ties between leaves of a cavity wall, opening and closing/slamming of doors and windows, roof movements.

Figures 12.7 and 12.8 are of a Georgian period house that has suffered a number of problems. Brickwork has been rebuilt at ground floor level and between the windows. Figure 12.9 shows a cracked lintel to an Edwardian house.

The BRE has issued guidance on the replacement of lintels (BRE, 1991) and retro-fitting of bed joint reinforcement (BRE, 2004).

Mortars may contribute to problems. Traditionally mortars were mixtures of sand and lime, with the component materials obtained locally. Such mixes generally accommodate movement quite well without failing. However if the sand has been obtained from a sea-side location it may be high in salts; if water for mixing was not clean, there may be impurities that might contribute to failure. However, problems related to sand/cement mortars are more common. A cement-rich mortar will have high load-bearing capability and durability, but is more prone to cracking; it is

Figure 12.6 Oversailing brickwork.

resistant; it does not *accommodate* movement. The Concrete Society (2005a) has issued guidance on the specification and use of mortars.

Renders have similar problems. Sometimes cracks in renders may be a manifestation of a co-related crack or shortcoming in the substrate, typically blockwork or brickwork. Often cracks appear where there is a change in substrate, for instance in proximity to a timber post or steel stanchion beneath, even where expanded metal has been provided to 'bridge' or moderate the change. The render specification should be (have been) chosen with the substrate in mind to bond to it and move with it as it expands and contracts with changes in humidity and temperature. The surface texture of rendering also affects its performance. Smooth renders are more likely to craze (see below).

The presence of one crack may be indicative of a one-off defect or deficiency, but it might not – there may be systemic problems, and a systematic survey should be carried out. This will enable the extent of problems to be assessed and appropriate remedial measures to be determined (BRE, 1983e, 1995c, 1995d; BSI, 2005b; Concrete Society, 2005b).

Crazing

This is less prevalent than cracking; it could be considered perhaps to be a subset of cracking. Crazing typically appears on smooth rendered external walls and in a pattern somewhat like that of crazy paving. It is not generally dangerous other than that the render may be inadequately

Figure 12.7 A Georgian house with problems related to its brickwork structure.

bonded to the blockwork or brickwork beneath, and may therefore be a danger to passers-by. Loose or hollow render can be identified by gently tapping the surface with a hammer, listening for a ringing sound. It should be removed, cut back to a line and replaced appropriately. Matching materials, including constituents, colour and texture of finish is difficult, so it is worth giving regard to the location, extent and visibility of the 'patch' and considering a larger job, including redecoration or cladding.

Spalling or delamination

Some materials have a tendency to lose some of their surface from time to time. For instance, some bricks may shed some of their surface, often irregularly, and some stones may erode. This can be caused by frost action or be related to the presence, especially in bricks, of sulphates or salts. When an external wall is wetted by rain it will tend to absorb it, and then dry out over a period. If however that dry period does not follow but the wall is subjected to below-freezing temperatures, the wall will freeze to some depth. Frozen water takes up more volume than when unfrozen and that expansion can result in the surface being 'blown off', or spalling. In the case of brickwork this tends to affect individual bricks differently, due to variations in the clay, their pressing, etc., so the effect is often patchy. With stone, there is more consistency through the

Figure 12.8 A detail of brickwork rebuilding related to subsidence.

stone beds, and what tends to happen is that a whole layer of stone will be shed (delamination), especially if the stone has been face-bedded.

Both brickwork and stonework are vulnerable to degradation by salt attack. Again when wet, salts tend to migrate to the mortar where, reacting with the cement, it manifests as a white powder on the surface and cracking of mortar beds. Sometimes mortar joints expand and spall off adjacent weak surfaces of bricks and stones. If inappropriate materials have been used there is little to be done to remedy the situation, but to try to ameliorate it. Details can perhaps be amended to keep the building (or at least its more vulnerable parts) drier, by shedding water higher up, providing cappings and copings, door hoods, mouldings or canopies, extended window cills, etc. In cavity walls, individual bricks or stones can be cut out and turned around, but this is expensive and can be just as unattractive until such time as the newly exposed surface has weathered in to match the surrounding wall.

In industrial areas, walls may be subject to acid rain, where pollutants fall in suspension in the precipitation (rain, snow, etc.) and grimey, sooty deposits may accumulate on wet walls. This can be thought unattractive (the 'dark satanic mills' that the poet Blake referred to in the hymn *Jerusalem*). There are various cleaning methods that could be applied, including washing with water or weak acid solutions or grit-blasting (BRE, 2000a, 2000b; de Vekey, 2001, 2008). Some of these techniques are rather aggressive and professional advice should be sought. They do not *solve* the problem; they deal with the symptoms.

Figure 12.9   A cracked lintel to an Edwardian house.

Verticality or bulging

Walls may occasionally be out of vertical, whether wholly or in part. Where foundations have subsided, the walls built off them may be out of true. Other possible causes, or contributions, may be floors or roofs which have deflected or otherwise moved and pushed out or pulled in or over walls into or on to which they have been built, or walls that are taking excessive loads. This could happen where the use of a building is changed and its loadings increased, for instance changing a house to office use whether in whole or in part. Roofs tend to spread over time as joints or timber members weaken, or when roof coverings are replaced unthinkingly with heavier modern materials, and can result in eaves being pushed out. This can happen more readily where walls are less well restrained, for instance at window lintels. Walls may also expand and bulge under the effect of frost or salts, especially where walls are wet, for instance from rising damp or leaking gutters. Regular gutter clearing should be a routine part of a maintenance plan.

Dampness

This may be manifested on external walls by a change in colour, either over time, or between parts of the wall, with a certain dullness in the damp areas. In particularly severe cases the wall may grow moss or lichens and take on a grey or greenish hue. Sometimes there may be plants (such as ferns or buddleia) growing in the damp wall areas. These can be common manifestations where gutterings or downpipes are inadequate and blocked, overflowing, at high level, unobserved, unattended and uncared for.

Damp external walls can also be observed internally. Sometimes, especially on upper floors or where the exterior of walls is obscured, hidden from view or not easily accessible, it can be seen more readily internally. It is most likely to be manifest as discoloured, stained or peeling wallpaper or other decoration such as paint.

Dampness may be caused by a range of factors (BRE, 1997a; Burkinshaw & Parrett, 2003; Trotman *et al.*, 2004). Possible factors include rising damp from the ground around or under the building (BRE, 1989, 1997b; British Standards Institution (BSI), 2005a; Trotman, 2007), penetrating damp (e.g. inadequate wall constructions or broken gutters) (Newman, 1988) and surface condensation. There may also be interstitial condensation within external walls and unobserved until manifested in damp conditions coming through on to wall surfaces. Condensation can be a complicated issue, affected by the combination of heating, insulation and ventilation (BRE, 1982, 1993, 1995a, 1997c).

At low level, dampness may be an indicator of lack or inadequacy of a DPC. Early DPCs took the form of a layer (or two or three) of slate, either vertically against the exterior of the wall, or horizontally in a bed joint. The advantage of multiple layers is to give added resistance and cover at joints. Slate is a relatively impervious material; a weakness, however, is its tendency to crack if the courses above or below it are subjected to significant movement, for instance from subsidence.

Sometimes DPCs have taken the form of a course (or two or three) of engineering brick selected for its resistance to the passage of water. This is generally quite effective. However moisture may travel through mortar joints if they are permeable, so the resistance of the wall as a whole is not always as good as hoped for.

More recent DPC materials, bitumen-based, hessian-reinforced or plastics, can tear and split under shear, or they may be incomplete at junctions or changes in level.

Oliver *et al.* (1997) give detailed coverage of dampness issues; Burkinshaw (2008) is good on remedial measures.

Rectification work

Observation, 'tell-tales' and photographs over a period of time can be very useful to help determine whether there is a seasonal or diurnal pattern or a declining trend, provided good records have been made. As indicated earlier, a full picture must be gained of the problem, including the consideration of a range of possible solutions, before rectification work can be decided upon or implemented.

Repairing cracks

Small and narrow cracks or crazing often warrant no remedial work. Where cracks are widening, however, they must be kept under review, and their causes identified and neutralised so as to halt further deterioration of the situation. This may involve underpinning of an inadequate foundation or stabilisation of a floor or roof structure.

Cracks of between 2 and 5 mm are capable of being filled, with the proviso that the situation has been stabilised. A weak cement–lime–sand mortar (maybe 1:3:12) is probably best to allow for some future movement. This will not, however, provide resistance to the passage of moisture, being low in cement. Where surrounding mortar is rich in cement, a 1:1:6 mixture may be better. It may be wished to rake out mortar from adjacent wall areas, so as to blend in, but care must be taken not to damage brick or stone arrises (edges) by over-vigorous raking out.

'Stitching' of wider cracks may involve the taking down and rebuilding of areas of brickwork or blockwork or stonework. The rebuilt work will almost certainly stand out visually from the original adjacent to it. If that is visually unacceptable, it may be that rendering or cladding in some way may be considered, or painting the elevation(s). If such solutions are proposed they must be fully considered so that further problems are not introduced inadvertently. For instance, new cavities may be introduced, providing voids through which fire may spread. Taking the opportunity to add insulation may result in dewpoint temperatures being achieved with subsequent condensation within wall structures. Inappropriate rendering on to poor substrates may result in further problems.

Underpinning

Underpinning may be used to stabilise the foundations on which walls sit. This was discussed in the previous chapter. The aim will normally be to hold a wall in the position to which it has fallen and for the associated crack(s) which have arisen to be filled, if not too great. It may be that the section of disrupted wall can be taken down and rebuilt so as to fill the gap, to recreate the impression of a unified piece of wall. The issue of satisfactorily blending will again be relevant; it may be acceptable for the rebuilt section of wall to look like it has been rebuilt!

Wall tie replacement

This is a specialist activity. Normally the existing wall ties have to be removed, not only because they have failed and are no longer helping hold the wall together, but because they will continue to rust and expand further, exacerbating problems. This rust expansion will have cracked and weakened bed joints in the brickwork or blockwork so the integrity of the wall has already been compromised (BRE, 1983d, 1996). Replacement wall ties are typically introduced by drilling into the wall from the outside and fixed in position by mechanical tightening and epoxy resins. There are also systems that tie the two leaves of the cavity wall with polyurethane-based materials, but these can be problematical in relation to water ingress and its transfer to the inner leaf through cracks in the cavity fill material.

Silicones

It may be thought that a way of reducing moisture-related problems in a wall is to provide a barrier to its entry from the air around. Some firms offer silicone-based treatments brush-applied to the exterior of a wall. The silicones are intended to penetrate pores of the bricks and thereby to provide resistance to the passage of moisture. However, coverage may not be complete and as a result moisture may be concentrated at the places where the application has failed. If the wall is wet when the silicone is applied, trapped moisture will be drawn to the untreated surface and its exit there may bring forth salts from the wall and internal finishes, resulting in undesirable white deposits on the wall internally.

Movement joints

Where expansion or contraction of a wall has been excessive, it may be desirable to introduce movement joints to accommodate future movements. Organisations like the Brick Development Association provide advice on the widths and spacings of such joints. Sawing with a brick saw may be a possibility (it will be very noisy and dusty, so proper provision and supervision will be essential), although it will be difficult to make this look neat. Alternatively, areas of brickwork may be taken down and rebuilt to incorporate a newly formed movement joint – in which case attention will again need to be given to the appearance of the finished work. Such joints would normally be placed so as to coincide with the points of weakness in the structure, where cracks have probably formed, for instance at and between door and window jambs.

Render repairs

It is difficult to repair render without it being very apparent. If it is decided to not replace render wholesale it may be as well to allow for a complete redecoration once the new work has fully dried out. The specifications will need to spell out the constituent materials and their proportions, and the number and thickness of coats and their timing and texture. Preparation, priming and bonding are key considerations. Render, properly done, is not a cheap alternative to face brickwork.

In some areas of Britain there are timber-framed buildings with infill between the timbers with lime-washed, horsehair-reinforced plaster on timber laths. This is not, strictly speaking, render but much the same considerations apply. Earth-based constructions such as cobb have sinilar considerations. These traditional or vernacular materials are now rather specialist and should be dealt with by people with those craft skills and experience.

Concrete

Concrete structures and finishes are worthy of special consideration. Concrete is often used where its structural and weather-resisting performances are important. Wherever the surveyor has doubts about the extent to which the structural integrity of a building or component part may be at risk, a structural engineer should be consulted. Concrete frames or panels may manifest problems like:

- cracking or crazing;
- spalling;
- staining;
- dislocation of elements.

Reinforced concrete with its combination of concrete and steel is particularly prone to problems. Cracks visible on the surface of concrete elements are almost invariably an indication of problems related to the steel reinforcement beneath. Often the reinforcement is poorly placed and too close to the surface. When moisture penetrates the concrete to a depth where it reaches the reinforcement in the presence of oxygen, for instance a crack, the steel will oxidise, rust and expand, forcing off the concrete.

There has been growing interest in the monitoring of concrete structures and their repair (e.g. Department of Trade and Industry (DTI), 2007; Matthews, 2007; Tilly & Jacobs, 2007; Construction Industry Research and Information Association (CIRIA), 2008).

Some concrete mixes were specified, especially in the 1960s and 1970s, to incorporate admixtures of various kinds. These were intended to facilitate the pouring and curing of concrete in adverse conditions, such as during cold weather. However the longer-term effects of some of these concrete mixes were not fully understood, and problems have consequently arisen with some of them. Investigations of original documentation may throw light on some situations, but often changes will have been introduced between the design and construction stages and such variances will not necessarily be well recorded. Such problems may be found in both *in situ* and precast concrete. It may be necessary to take core samples to be sent away for laboratory testing to ascertain the nature of the concrete and to help form a view of what, if anything, should be done.

Figure 12.10 Concrete repairs are hard to hide.

Cracks and spalling may not necessarily be indicators of a terminal condition, but a full and professional consideration is required because lay people may need to be reassured.

Concrete shows its weathering badly with streaks in varying shades of grey. Brown staining may be a sign that water has reached reinforcement beneath, but it might not be that. Some aggregates used in concrete can have iron pyrites that may cause brown staining.

Generally it is thought unavoidable to 'live with' stained concrete rather than try to clean it up. Spalled concrete however, normally attracts repair in the form of small pockets of patched concrete, in order to provide cover to the exposed reinforcement and/or to restore the surface, but the new work is very apparent (Figure 12.10)! As with poor brickwork or blockwork, consideration may be given to covering the concrete somehow, perhaps with sheet cladding of some kind. In the 1990s quite a lot of over-cladding was done, to protect vulnerable concrete structures beneath and to slow their deterioration. Often these were able also, and by design, to secure a substantial makeover in terms of a transformation of the image of the building.

People in Britain at least (and not only in Britain) have found it hard to learn to love concrete. It has often been used as a cheap, mass material, lacking interesting detail and poorly looked after – a feature of high-density urban areas and the 'brutalist' era. It is therefore appropriate to consider concrete repairs and alternatives against that background.

The 1960s and 1970s were characterised in Britain by wholesale clearance of slums and the building of tower blocks to replace them. There was a massive public housing programme, and quantity was more important than quality. The UK government, in its quest for speed and productivity, promoted large-scale 'industrialised building'. This built on the immediate post-war developments of non-traditional aluminium, steel and concrete prefabricated housing. Much of the new building took the form of large panel system (LPS) concrete housing. On the whole these massive structures were unsuccessful socially (and economically). Many of these LPS and the earlier PRC (prefabricated reinforced concrete) designs developed systemic failures. The BRE has accumulated and published a huge amount of data on these, and several firms of structural engineers and contractors have developed expertise in how best to deal with them. It is as well to make use of that experience and, should you be asked to deal with such estates, you would be well advised to propose use of that experience.

Claddings

By and large, claddings are materials used as non-structural components to provide a weather-resisting 'skin' to a building. Common forms would include timber boarding (or metal alternatives), slate- or tile-hanging, and metal or concrete panels. Other materials used have included fibreglass, asbestos–cement and various ceramics.

Most timber used in building is softwood – this is relatively quickly grown, readily available, cheap and versatile. Some situations warrant the use of hardwoods, which are generally less available and more expensive. There is a common misconception that hardwoods are more durable than softwoods; generally they are but there are exceptions, so it is important to select species with care. Timber, an organic material, deteriorates, and will need regular maintenance with repair and/or replacement from time to time (BRE, 1983c, 1995b). Earlier chapters have discussed the need to take a considered view about day-to-day repair needs and planning programmes of repair and replacement works. Timber claddings (and windows and doors) are areas of work that are likely to feature in such programmes, as failures are often systemic and repetitive (as opposed to foundation or brickwork failures which tend to be much more one-off).

Slate-hanging is more common in areas where slate is more common – such as Wales and the south-west of England. Tile-hanging is more common in Kent, Surrey and Sussex. Often the tile shapes are interesting (arrowhead, bull-nose, club and fishtail being four) which is visually exciting but can make for difficulties of availability of replacements.

Timber claddings generally fail because of warping or splitting of the timber or because the fixings (usually nails) have failed. Some timbers have acids within them that may react with fixing materials, though this is relatively rare in these circumstances. The effect of rain, wind, ice and sun over time is generally enough. Normally, materials are cheap enough to suggest replacement of failed boards rather than piecing in of sections of a board. Often the inherent variation of wood is such as to feel that replacement of individual boards as and when required is acceptable, rather than feeling that whole elevations should be replaced at the same time.

With slate-hanging or tile-hanging, issues are likely to relate to broken or missing individual slates or tiles (or small areas) or may be linked to dislocation or cracking related to subsidence or movement in the building for some other reason. Clearly in those situations enough investigation of the surrounding context needs to be made to enable decisions to be taken about the way forward. A common approach is to 'cannibalise' some areas in order to provide old slates or tiles that will match in with the existing in the retained areas. Often these materials are features of an area thought worthy of protection by listed building or conservation area status, so consultation

Figure 12.11 Stained cladding panels.

with the conservation officer of the local planning authority, and approval, will almost certainly be required.

Many buildings constructed since World War II are framed structures with panels to provide the external skin. Many such panels were made of asbestos–cement. Many of those have already been replaced but there will be others still in use. Asbestos has been recognised as a dangerous material and in Britain and many other countries there are stringent controls and licences are required to work with it. It is important to be able to recognise in any situation the possibility that asbestos-containing materials may be present. Once such presence is suspected then the situation must be professionally inspected and evaluated. This is a specialist activity. The mere suspicion of the presence of asbestos or discussion of such is likely to raise concerns and anxieties, so it is important to be sensitive and to take this very seriously.

The cost of removal of asbestos-containing materials, including the costs of disruption (financial and operational), is likely to be so great as to make it worth having a major review of the long-term prognosis for the use and configuration of the building, including considering the possibility of demolition.

Figure 12.11 shows staining of (and minor damage to) plain-finish cladding panels; the panels are especially damp in those parts where rainwater has been concentrated. The brickwork beneath is also stained due to water shed from the panels.

Windows and external doors

These must be considered both as units in themselves and as openings in external walls; these will be considered together here. Typical problems to deal with include:

■ doors and windows not shutting (or opening);
■ associated draughts;
■ glazing-related issues;
■ security and vandalism.

Contributory factors can be:

■ decay and deterioration, such as rot and rust;
■ distortion: twists and shakes;
■ wear and tear;
■ poor specification or installation.

Timber is the most extensively used material for doors and windows in domestic properties and coated steel or aluminium is most often used in the commercial world. As discussed briefly elsewhere, timber can be hardwood or more commonly softwood. Often door and window cills and thresholds will be in hardwood even though the rest of the unit and its frame may be in softwood. This is intended to provide a more hard-wearing timber in the more vulnerable parts. Timber has the advantage that it can be fairly readily worked and adjusted on site – the unit can be removed and rehung, sawn and planed, have pieces removed and/or added, surfaces filled, sanded and redecorated as required (BRE, 1997d, 1997e). Better still, decay can be prevented or at least arrested by appropriate considerations (BRE, 1983a, 1983b). Metal units may be harder to adjust, other than through their ironmongery, but may be considered more stable and therefore less likely to require such adjustment.

Figure 12.12 shows extensive decay to a timber window; replacement with new windows to a higher specification would be appropriate. Figure 12.13 shows 'piecing in' to a window transom – however decay has continued beyond, suggesting that the repair did not go far enough. Figure 12.14 is a large metal door unit. Substantial rusting has taken place and it is unlikely that redecoration, even with thorough preparation, will be worthwhile, especially as the unit is single-glazed; replacement should be considered if the building has a life ahead, which it may not.

Swelling and sticking

Timber doors or windows may stick due to swelling. This is most likely to be caused by moisture. Timber shrinks and swells in relation to changes in its moisture content. External doors and windows, by nature of their function, are at the interface of external and internal environments, and are intended to mediate between the one and the other. Both external and internal conditions also change, seasonally and through the day. Sometimes – in the UK at least – these elements may be subjected to the weather of four seasons in one day.

Painting and varnishing of doors and windows was intended, in addition to giving a finish of good appearance, to provide protection to the timber beneath. (Decoration is discussed further below.) Where the decoration is incomplete, for instance due to damage or wear, or poor work in the first place, this provides access for moisture to the material it should have

Figure 12.12 A rotted window.

Figure 12.13 A pieced-in transom, with more rot beyond.

been protecting. Indeed, especially if combined with poor detailing that guides water to it, there will be concentrations of high levels of moisture at these places, enhancing the risk of rot or rust.

Where a small area of a door or window is affected by rot it may make sense to just cut out the affected area (and a little beyond to ensure all the affected material is removed) and to 'piece in' a matching section; then redecorate etc. It will be important of course to identify the cause of the wear or damage and the source of the moisture in order to address and resolve those too. A

Figure 12.14 A rusted metal door unit.

larger area, or a systemic one seen to be repeated in many similar situations, of course demands a consideration of the 'bigger picture' in determining the appropriate action.

What is generally not advised is a simple easing – a cutting or planing of the door or window to fit – without considering the unit's context and its subsequent performance. For instance, if the unit has become exceptionally wet due to unusually high rainfall it may make sense to wait for it to dry out before determining action. Material once removed is not easily restored; and there may be gaps in the summer when all is dry.

The ease of opening of doors and windows may be improved by an occasional lubrication or 'greasing' of closing faces by rubbing with a candle; and hinges should be periodically more formally lubricated to inhibit rusting. Sometimes ironmongery rusts and contributes to sticking, where for instance decoration is incomplete or screws are neither galvanised nor sherardised. In these cases, removal and replacement of the 'offending' components should be carried out.

Sticking of doors or windows may also be because parts of the wall structure within which they are located were poorly designed and/or constructed and/or have moved. That movement may be for any of the reasons discussed earlier in this chapter and manifested in any of the ways indicated there. The sticking of a door or window may be a pointer to such a problem, and the possibility should be considered before putting in hand any adjustment to the door or window.

Shrinkage and gaps

Doors and windows that have been poorly stored before installation may be wet in part (or more fully). They may therefore be swollen and dimensionally unstable at the time of installation and liable to shrink as they and the building dry out over time. Internal environmental conditions may be warm and dry. Shrinkage of doors and windows may result in gaps between the opening part and the frame, which may be unacceptably large. These gaps may be unsightly; they may allow draughts to enter; they may result in ironmongery not quite fitting correctly; doors may rattle in their openings. Similarly, frames may shrink; they may pull away from their locations in wall openings, loosening fixings, opening up gaps from plaster, tearing mastic or silicone joint fillers.

Cracks may open up in the timber itself, whether as a check (a fissure not extending through the thickness of the timber) or as a split passing through the thickness. While these fissures may open and close with the timber's changing moisture content, they will not disappear! They may start from shakes in the original timber, which confirms the need for good timber specification and selection at the outset, for there is little that can be done to redeem the situation short of replacement. A short-term filling may suffice for a little while but is unlikely to be satisfactory in the long run. A similar problem may occur with knots in the timber – drying out is likely to give rise to the knot falling out when it shrinks.

Distortion

When timber shrinks it does not do so consistently. The grain of the wood, the growth of the tree and its 'conversion' firstly into a piece of timber and then into a door or window or frame mean that it is not internally consistent. Thus as it dries out some parts shrink more, or more quickly, than other parts of the same member, causing the timber to distort. Twisting, cupping, bow, spring and other distortions can all cause problems with opening lights or door leaves ceasing to fit snugly into their frames. There is no treatment that is likely to return a distorted timber to its original shape; although easing may provide some respite for a short period, replacement is the only solution.

Timber should be properly seasoned, or kiln-dried to a moisture content similar to that it will inhabit in the building before being used in building components. Distorted timber should be rejected. Timber members which are small in section are generally more liable to distort than more substantial sections and this should be borne in mind when designing, detailing or selecting doors and windows. A 'chunkier' window may be less visually appealing and admit less daylight, but it is likely to be more durable. Frames that feel flimsy are quite likely to be problematic both in their fixing and in their lack of longevity.

Weathering and detailing

Unlike the external walls within which they are fitted, which may be expected to deal with precipitation by absorbing some of it (like an overcoat), doors and windows are expected to shed rainwater. This they do by having generally impervious surfaces and by designs that 'fall' the water away from the building. They tend to have high concentrations of water at their cills as water falls down the door or window, especially where there are large glazed areas. Window beadings of small section are especially vulnerable. Horizontal members are usually provided with slopes to their upper surfaces, overhangs to cills to extend out beyond the plane of the wall below, and throatings on their undersides so that water does not run back on the lower surface and on to or into the building thereby. The absence of any of these features, or poor provision of them – for instance shallow cills – should be considered as possible contributory factors to either poor performance of a door or window in itself, or in relation to the openings in which they fit (or do not quite fit!).

Door and window openings may in many ways be considered points of weakness in the external envelope; they should therefore be accorded appropriate and commensurate attention. The principal weakness or vulnerability is at the junction. Mastics and fillers of various kinds saw a large increase in use in the latter part of the twentieth century. They may be good at filling unsightly gaps; their formulation, particularly in terms of their elasticity and adhesive qualities may have advanced greatly, but they should not be expected to be the main element of weather-resistance. Where they have failed, a full consideration of the junction should be made rather than just recaulking.

Metal windows

While timber may be the traditional material for windows, the 1930s particularly saw the introduction of metal windows on a large scale. Crittall windows became a feature of the modern movement that swept across Britain. Mass manufactured in Braintree, Essex, they were also a feature of the company town of Silver End nearby. A great advantage of the steel window is its small frame section, allowing the new houses and offices to be bathed in light internally. The 'suntrap' window is a product of that time. Steel rusts; the Crittall window dealt with this by adding galvanising. This provides a layer of non-rusting zinc to the surface of the steel section to resist corrosion. Its effectiveness depends upon its completeness. So missed or damaged areas are prone to rusting. Early versions had more imperfections.

Where steel windows are rusting it may be possible to rub down the affected area and to paint on inhibiting treatments, but the long-term solution is replacement. Because of their slim section, it is also possible that metal windows may be installed with an unintended twist. This may be unnoticeable for a time but come to light as a result of perhaps some subsequent movement in the building and perhaps cracked glazing where stresses have built up. Rusting may also have this effect. The Steel Window Association (SWA) offers guidance on the aftercare and maintenance of steel windows (SWA, 2003).

Partly in response to problems with steel windows, the 1960s saw the introduction of aluminium windows. While generally steel windows were installed as either side-hung or top-hung casements or horizontally-pivoted, these aluminium windows were more commonly provided as horizontal- or vertical-sliders (the latter, in some ways, mimicking the double-hung sash windows beloved of the Georgian, Victorian and Edwardian periods). Early models were generally of plain mill finish, prone to pitting and white 'furring' due to oxidation. These problems were more cosmetic than structural but contributed to the decline in specification of aluminium windows. More recent developments have brought anodising and other finishes to improve appearance. Again, problems tend to relate to slenderness of the section and movement of the structure. These are commonly exacerbated by the nature of the building into which the units have been designed. Tall structures tend to have accessibility issues when it comes to questions of repair and replacement. It is sensible in these situations to include for periodic inspections perhaps as part of a cleaning regime; the consequence and cost of failure could be high.

Figure 12.15   Poor detail to a bay window roof.

Bay windows

These can be very troublesome elements. They are often provided to give visual interest and variety externally, and more space internally, but their constructions are not always sound. They are prevalent in domestic properties from the Victorian period and since. They are almost like little structures on their own but without the attention they should have had at the design and construction stages. They may be built on inadequate foundations, be poorly tied into the main external walls, have poor or no flashings where the roof abuts the wall and have inadequate rainwater collection and disposal, for instance. The walls above the opening for the bay window may be poorly supported. When problems present, they demand thorough inspection of each of the component parts. Sometimes it has been deemed better to remove the troublesome bay; this invariably leaves a very difficult void to fill and the visual impact of the 'infilling' is often degrading of the neighbourhood. Figure 12.15 shows a detail of a felted flat roof to a bay window where an attempt has been made to remove some of the lead flashing; a combination of poor specification and vandalism.

Doors

The 'front door' is often a significant part of the outward impression or image of what is behind it, whether domestic or commercial. Large, imposing houses, hotels and banks for instance have traditionally had large, imposing doors and doorways at their main entrances; they should not be removed or changed lightly. It may be necessary therefore to consider a more conservative (in the sense both of conservation and of being less radical) approach to the repair and restoration rather than replacement of such a door, with all its historical detail. In the event that replacement may be necessary, an accurate replication will be important.

The importance attached to doors is also attested to by the attention given to replacing the front door as a very clear statement of ownership when former council tenants throughout the

UK bought their houses under the right-to-buy initiatives of the 1980s. The new doors may not have been of superior constructional quality but they had the definite merit of not looking 'council'.

The Door and Shutter Manufacturers' Association (DSMA) has prepared a code of practice for the maintenance of industrial and commercial doors (DSMA, 2003).

Glass and glazing

Necessarily all windows and many doors are glazed (although originally 'window' meant wind-hole and was seen as primarily a ventilation device). While the most common maintenance activity in relation to glass may be its cleaning (and in some buildings that is a very considerable activity) there is need to do more from time to time. Glass is brittle, so by and large when it fails in performance it is generally because it has cracked or been smashed. The former may occur due to movement of the building or some human involvement, the latter almost certainly human. Usually broken glazing is replaced with glass like-for-like, though occasionally polycarbonate or similar sheet may be used for better resistance to breakage – it is inclined however to discolouration and loss of transparency over time.

Occasionally windows may be blown in or sucked out under extreme weather conditions, in which case the nature of the fixing will need to be seriously reviewed before replacement. In high buildings, building movements can be quite appreciable; designs need to take this into account. The mirrored-glass façade of the then new John Hancock Tower in Boston suffered fracture and loss of many glazed panels in the 1970s, necessitating the provision of acres of plywood to prevent further dangerous falls of material. This will have caused the loss too of millions of dollars in rectification works; the cause of the fault remains unpublished, apparently a proviso of the settlement of the lawsuit.

Work on highly glazed buildings requires particular care; BRE, CIRIA and the Glass and Glazing Association (GGA) have published guidance (Keiller et al., 2005; Ridal et al., 2005; GGA, 2007).

Double and triple glazing; replacement doors and windows; PVCu

Whenever single-glazed windows fail, their replacement with double- or triple-glazed windows must be considered. UK Building Regulations now require that replacement windows meet today's standards – like-for-like replacements are no longer acceptable. It makes sense to see such upgrading as part of a strategic overview of a building because the improvement of a single window may result in condensation problems elsewhere in the building – a holistic view is required. The saving in energy use over a period of years should be sufficient in any case to justify wholesale replacement of single-glazed windows. Unfortunately it is rarely possible to replace single-glazing with double-glazed units (let alone triple) as the rebates in the frame are generally too shallow and the frames too flimsy. It is therefore necessary to expect to take out all the old windows and replace them with new. This seems poor from a sustainability point of view where one might hope to reuse or recycle, but we must consider the overall gain to be worth that loss.

Double- and triple-glazed units are also appreciably heavier than those they replace, so more durable ironmongery and fixings are required too. This may be a factor in the sometimes disappointing performance of PVCu windows and doors. Studies have shown (e.g. Berry, 2000) that ironmongery failures on PVCu windows, and especially doors, can be quite high, perhaps as a result of poor specification or poor performance. Often multiple-glazed units are designed to be

Figure 12.16 A cracked precast concrete cill.

fixed lights and not openable, airtight, and thus dependent upon provision of an air-conditioning system. This may be perhaps an unintended consequence of well meant improvements in regulations, but that is beyond the scope of this book.

Cills, heads, reveals and canopies

These are often neglected areas but very prone to damage. Typically, cills will be of canted or special-shaped bricks, precast concrete units or tiles. In the 1950s there was a fashion for concrete window surrounds – a set of precast units comprising cill, head and reveals projecting beyond the external face of the external wall. Many of these have performed badly, with spalling where cracks have allowed rainwater to reach the reinforcement beneath, blowing off the concrete as the steel has rusted and expanded. Figure 12.16 shows a cracked concrete cill and Figure 12.17 shows a more advanced example – a spalled concrete cill on the same building.

External doors provide opportunity for water ingress. Canopies enable callers to shelter while waiting to gain entry, but are themselves exposed to the weather. Concrete canopies, like cills, are rich sources of problems related to cracking and spalling, especially of their bracket supports. Figure 12.18 shows an example of a lead-covered canopy supported on decorative wrought iron brackets.

Decorations

Often the most obvious sign of maintenance needs on the exterior of a building is the poor state of its decoration, most commonly paint flaking and peeling, perhaps showing bare wood beneath. As with any other aspect, it is important to consider underlying factors before launching into a wholesale redecoration. Peeling paint may be a result of lack of redecoration for many years; it may also be an indication of dampness in the material beneath, which may arise

Figure 12.17 A spalled precast concrete cill.

Figure 12.18 A decorative door canopy.

from rising damp, or a leak, or a poor eaves detail, or any number of possible contributory factors.

It is common, when budgets are tight, to cut down the number of coats in a painting specification, or to skimp on the preparation work. There is no substitute for providing a good substrate; if this is not done, subsequent layers may not adhere as they should, and the finished product will not last as long as it should if done properly.

Damp materials should not be painted upon. The cause(s) of the dampness should be identified and eradicated, and the wall dried out before applying a fully thought out decoration specification, if indeed that is the best solution. Some decorative finishes intended to resist the passage of water and thus to shed rain like a raincoat may, however, result in moisture being 'locked in' if they are impervious.

Some timbers may have become so rotted by the absence of paint for such a long time that they need to be replaced in part or in whole. This should be done as part of a thorough preparation. This timber may be provided pre-primed, in which case it is important to specify that all cut ends should be primed on site, and to check that that is done, or water will be taken up very readily by the open end-grain at the cut.

Where existing paint is peeling, it must be stripped off back to the bare wood, primed and repainted in the requisite number of coats – typically two undercoats and one gloss coat. It is normal to specify that successive coats should be of different shades so as to provide ready evidence that the correct number of coats has been applied. It is perhaps needless to say that incomplete preparation, for instance leaving some of the old paint on while in other places the wood is bare of paint, will result in an unsatisfactory end-product.

It may be thought that the solution to some unsightly wall may be to paint it; it is as well to think twice before doing that. Not only will a lot of preparation be needed, but once the wall has been painted, it will only be a matter of time before redecoration is needed. If that is not done expeditiously, the wall will again be seen to be in a poor decorative state – an unsatisfactory state of affairs, especially after the expenditure of the additional monies.

As budgets for decoration are so often cut when times are hard, it is well worth giving full consideration to the specification whenever possible of self-finished materials and components that will not require continual redecoration.

Summary

This chapter completes our consideration of the vertical weatherproofing exteriors of buildings. We now move on to deliberations on the parts of the building most likely to give rise to problems of water ingress – roofs.

References

Berry, S. (2000) *Maintenance of PVCu Double Glazing: Choices for Local Authorities*. Unpublished MSc thesis, Oxford Brookes University.

Bonshor, R.B. & Bonshor, L.L. (1996) *Cracking in Buildings*. BRE, Watford.

Brick Development Association (1986) *Design Note 10: Designing for Movement in Brickwork*. BDA, Windsor.

British Standards Institution (2005a) *BS 6576: 2005 Code of Practice for Diagnosis of Rising Damp in Walls of Buildings and Installation of Chemical Damp Proof Courses*. BSI, Milton Keynes.

British Standards Institution (2005b) *BS EN 13914-1: 2005 Design, Preparation and Application of External Rendering and Internal Plastering: Part 1: External rendering*. BSI, Milton Keynes.

Building Research Establishment (1982) *External Walls: Reducing the Risk from Interstitial Condensation: Defect Action Sheet 6*. BRE, Watford.

Building Research Establishment (1983a) *Wood Windows: Arresting Decay: Defect Action Sheet 13*. BRE, Watford.

Building Research Establishment (1983b) *Wood Windows: Preventing Decay: Defect Action Sheet 14*. BRE, Watford.

Building Research Establishment (1983c) *Walls and Cladding: Remedying Mould Growth: Defect Action Sheet 16*. BRE, Watford.

Building Research Establishment (1983d) *External Masonry Cavity Walls: Wall Tie Replacement: Defect Action Sheet 21*. BRE, Watford.

Building Research Establishment (1983e) *External Walls: Rendering – Resisting Rain Penetration: Defect Action Sheet 37*. BRE, Watford.

Building Research Establishment (1985) *External Masonry Walls: Vertical Joints for Thermal Movements: Defect Action Sheet 18*. BRE, Watford.

Building Research Establishment (1989) *Rising Damp in Walls: Diagnosis and Treatment: Digest 245*. BRE, Watford.

Building Research Establishment (1991) *Repairing or Replacing Lintels (rev. 1): Good Building Guide 1*. BRE, Watford.

Building Research Establishment (1992) *Why do Buildings Crack?* Digest 361. BRE, Watford.

Building Research Establishment (1993) *Cavity Wall Insulation in Existing Housing: Good Practice Guide 26*. BRE, Watford.

Building Research Establishment (1995a) *Energy Efficient Refurbishment of Low Rise Cavity Wall Housing: Good Practice Guide 175*. BRE, Watford.

Building Research Establishment (1995b) *Maintaining Exterior Wood Finishes: Good Building Guide 22*. BRE, Watford.

Building Research Establishment (1995c) *Assessing External Rendering for Replacement or Repair: Good Building Guide 23*. BRE, Watford.

Building Research Establishment (1995d) *Repairing External Rendering: Good Building Guide 24*. BRE, Watford.

Building Research Establishment (1996) *Replacing Masonry Wall Ties: Good Repair Guide 4*. BRE, Watford.

Building Research Establishment (1997a) *Diagnosing the Causes of Dampness: Good Repair Guide 5*. BRE, Watford.

Building Research Establishment (1997b) *Treating Rising Damp in Houses: Good Repair Guide 6*. BRE, Watford.

Building Research Establishment (1997c) *Treating Condensation in Houses: Good Repair Guide 7*. BRE, Watford.

Building Research Establishment (1997d) *Repairing Timber Windows: Investigating Defects and Dealing with Water Leakage: Good Repair Guide 10: Part 1*. BRE, Watford.

Building Research Establishment (1997e) *Repairing Timber Windows: Dealing with Draughty Windows, Condensation in Sealed Units, Operating Problems and Deterioration in Frames: Good Repair Guide 10: Part 2*. BRE, Watford.

Building Research Establishment (2000a) *Cleaning External Walls of Buildings: Cleaning Methods: Good Repair Guide 27: Part 1*. BRE, Watford.

Building Research Establishment (2000b) *Cleaning External Walls of Buildings: Removing Dirt and Stains: Good Repair Guide 27: Part 2*. BRE, Watford.

Building Research Establishment (2004) *Retro-installation of Bed Joint Reinforcement in Masonry: Good Building Guide 62*. BRE, Watford.

Burkinshaw, R. (2008) *Remedying Damp*. RICS Books, London.

Burkinshaw, R. & Parrett, M. (2003) *Diagnosing Damp*. RICS Books, London.

Concrete Society (2005a) *Mortars for Masonry: Guidance on Specification, Types, Production and Use: Good Concrete Guide 4*. Concrete Society, Camberley.

Concrete Society (2005b) *Rendering: Defects and Remedial Measures: Concrete Advice 24*. Concrete Society, Camberley.

Construction Industry Research and Information Association (2008) *Intelligent Monitoring of Concrete Structures*. CIRIA, Bracknell.

de Vekey, R.C. (2001) *Principles of Masonry Conservation Management*, Digest 502. BRE, Watford.

de Vekey, R.C. (2008) *Conservation and Cleaning of Masonry, Digest 508: Part 1, Stonework*. BRE, Watford.

Department of Trade and Industry (2007) *Automated Monitoring of the Deterioration of Concrete Structures*. The Stationery Office, London.

Dickinson, P. & Thornton, N. (2004) *Cracking and Building Movement*. RICS Books, London.

Door and Shutter Manufacturers' Association (2003) *Code of Practice for the Repair and Maintenance of Industrial and Commercial Doors*. DSMA, Tamworth, Staffs.

Driscoll, R.M.C. & Crilly, M.S. (2000) *Subsidence Damage to Domestic Buildings*. BRE, Watford.

Glass and Glazing Association (2007) *Guide to Good Practice in the Specification, Installation and Use of Replacement Windows and Doors in England and Wales*. GGF, London.

Johnson, R.W. (2002) The significance of cracks in low-rise buildings. *Structural Survey*, **20**(5), 155–161.

Keiller, A., Walker, A., Ledbetter, S. & Wolmuth, W. (2005) *Guidance on Glazing at Height*. CIRIA, London.

Matthews, S. (2007) *Performance-based Intervention for Durable Concrete Repairs: Information Paper 9/007*. BRE, Watford.

Newman, A. (1988) *Rain Penetration Through Masonry Walls: Diagnosis and Remedial Measures*. BRE, Watford.

Oliver, A., Douglas, J. & Stirling, J. (1997) *Dampness in Buildings*. Blackwell Publishing, Oxford.

Ridal, J., Reid, J. & Garvin, S (2005) *Highly Glazed Buildings: Assessing and Managing the Risks*. BRE, Watford.

Steel Window Association (2003) *Aftercare and Maintenance of Steel Windows*. SWA, London.

Tilly, G.P. & Jacobs, J. (2007) *Concrete Repairs: Performance in Service and Current Practice*. BRE, Watford.

Trotman, P. (ed.) (2007) *Rising Damp in Walls: Diagnosis and Treatment*, Digest 245.BRE, Watford.

Trotman, P., Sanders, C. & Harrison, H. (2004) *Understanding Dampness: Effects, Causes, Diagnosis and Remedies*. BRE, Watford

13 Defect recognition and rectification

Chimneys, roofs and roofspaces, rainwater disposal

This is the fourth of six chapters addressing common defects in buildings; it completes our study of the building's exterior parts.

Chimneys

These are significant because they are often the most exposed part of a building, and because they pass through roofs and thereby provide a point of weakness in the weather-resistance of the external envelope. They are often substantial elements and sometimes of architectural interest or merit. Being at the top of the building and often high above the ground there can be difficulties in gaining access for inspection and repair.

There is a number of component parts to a chimney. The major part is the stack, which is generally of the same material as the external walls from which it often extends, most commonly brickwork or stonework, and occasionally rendered. This is a weakness. Chimney stacks suffer much the same defects as the external walls – cracking, spalling and distortion. Where stacks are constructed of the same material as the walls below, they are likely to incur the same damage, and, being more exposed, perhaps more so or sooner. They are also exposed to severe differences in temperature and moisture. On the inside, hot gases (from the fire below); on the outside, howling winds and perhaps heavy rain or snow or hot sun.

It is important to give chimneys serious attention. Because they are above eye level, defects are often not seen by passers-by and therefore may not to be reported as readily as problems related to the car park, building entrance or lifts! Because they at a high level they can be dangerous – falling masonry can kill. And problems may be worse than expected; a small crack observed from the ground may be larger when viewed close up and/or may be accompanied by a larger crack on a less readily visible face. The Building Research Establishment (BRE) has published guidance on surveying chimneys (BRE, 1990a).

Chimneys may distort due to differential conditions. It may be that one or two sides of a chimney receive a near constant onslaught of wind and or rain, while other sides are more protected. This may be exacerbated by problems from within. Repeated heating and cooling from fires lit and extinguished may contribute to cracking of internal linings which may in turn allow flue gases to penetrate the structure of the stack. Chemical reactions may take place in the structure, salts may migrate with moisture movements, and differential expansions are sometimes so great as to be readily noticeable as affected chimney stacks turn banana-shaped. Stacks that are cracked or distorted should be rebuilt. This will be expensive, but so too would be a repair, as work will almost certainly require substantial scaffolding and associated costs. BRE has produced guidance on repair, rebuilding and lining chimneys (BRE, 1989, 1990b, 1998b). It may be worth thinking about removal of stacks where fireplaces are no longer in use, but care needs to be taken to maintain ventilation that the chimney may provide and to have regard to architectural considerations.

Great care must be exercised when taking down and/or rebuilding chimneys. Material dropped from this height will almost certainly be capable of killing anyone upon whom it may fall. Material may also fall down inside the building down the flue; this may be material dislodged unnoticed from within the chimney below where the work is being carried out, including cement linings and loose bricks remaining from previous works. Fireplaces must be blocked up against falls of such materials. There may be quantities of soot also ready to fall with dire consequences.

Exposed stacks are also liable to erosion of mortar joints. This is readily observable as a kind of 'sawtooth' profile when viewing the stack on the diagonal against the sky. Repointing may be appropriate, but this will also be expensive and may be problematic. Raking out of the existing joints must be done carefully or brick arrises will be damaged; it must be done to a depth of at least 20 mm so that the new mortar will not fall out. Sometimes, contrary to the conclusion reached from the initial observation, the mortar is too hard and resistant to be readily raked out to the required depth; it is worth carrying out work to a sample chimney as a 'pilot' project.

Chimney heads are another consideration; they are often another point of detail, perhaps with bricks of another colour and grade, and maybe oversailing the stack. They are generally more vulnerable than the stack on which they sit, being even more exposed; often chimneys crack from the head downward. Where cracked, they must be rebuilt. It is often visually degrading to rebuild only part of a stack, and similarly it is important to replicate the architectural details as far as possible. A wholesale rebuilding provides opportunity for a simpler construction if that is thought desirable. A thorough investigation and review should be undertaken and senior management time allocated to the development of a policy in relation to repairs and replacements, such that operatives can be directed accordingly.

Chimney pots and the cement flaunching which should be holding them in position are also often cracked; they too should be replaced. Pots are often very decorative and all attempts should be made to retain them. Sometimes particular designs of pot have been installed with a view to improving the draw of a chimney where perhaps the original height was inadequate. With the demise of solid fuel heating this is no longer the consideration it once was.

Chimneys also have points of weakness where they pass though roofs. Rainwater absorbed by the brickwork or stonework of a chimney, or shed from its surface, could enter the roofspace below. The provision of effective flashings is important. What is appropriate depends on the situation. A chimney located at the ridge offers fewer problems than one mid-slope or at the eaves. Lead is the most common material for flashings, including stepped flashings and soakers, and this is generally a good choice, offering sufficient malleability to be dressed into difficult corners, so long as enough material was provided and the detail was well considered and well executed. This is a skilled operation. As so much time and money has to be allocated to gaining access, scaffolding, etc., and the price of failure also high, it is worth spending the time to do a good job.

More problematic chimneys as regards flashings are those mid-slope and at the eaves – this is because both require an apron at the back-slope where water will be coming down the roof-slope behind the chimney and could be directed into the building at that point! It is important to check that such flashings:

▥ exist;
▥ are fully supported (usually by timbers, which may be rotten);
▥ direct water away from the chimney;

Figure 13.1 Complicated roof and chimmey junction.

■ keep water on the roof surface and do not direct it on to sarking felt below (which may be incomplete and/or split);

■ are complete and not cracked or split or penetrated by fixings such as nails!

Figure 13.1 shows a chimney with complicated roof forms around it; it is important to ensure that flashings are complete.

It is always worth checking chimneys from within roofspaces, too, for related evidence of problems above (see below).

Pitched roofs

There are several component parts to consider with a pitched roof. It may be quite complicated (often more so than a flat roof, but with fewer problems!), so it is important to consider a wide range of possible issues. The configuration of a roof is often very much dictated by the shape of the building on plan beneath it; its materials are often determined by the building's age and location.

The most common pitched roof materials in the UK are:

■ slate;
■ clay tiles;
■ concrete tiles;
■ asbestos–cement – as tiles or sheets;
■ fibre–cement (superseding asbestos–cement);
■ metal sheet roofing.

Slates

Slate is a commonly occurring material in some parts of Britain, and it is generally referred to by the place from which it comes, for instance:

- Westmoreland green slate;
- Collyweston;
- Stonesfield;
- Delabole.

When used for roofing, these slates are generally obtained from stones hewn from quarries (or fields, as in Stonesfield) of varying size and thickness and sorted into broadly similar sizes for constructing the roof. Stonesfield slates were fashioned by arranging the stones on the field surface and allowing the winter frosts to create the slates by delamination. Roofs of stone slates tend to have diminishing courses with larger slates at the eaves. This has the advantages of:

- using the larger and heavier slates at the foot of the roof, thus requiring less lifting of the heavier weights;
- placing the heavier weight where the structure can best take it;
- offering fewer joints where there is the greater concentration of rainwater, at the bottom of the roof-slope;
- enabling slates of all sizes to be used.

There are traditional names allocated to different sizes of slate (see table).

Name	Size of slate (inches)	Size of slate (mm)
Empress	26×16	660×406
Princess	24×14	610×356
Duchess	24×12	610×305
Marchioness	22×10	559×254
Countess	20×10	508×254
Viscountess	18×9	457×229
Lady	16×8	406×203

The 'heyday' for slate came with the industrial revolution and the growth of the railway network. This saw Welsh slate take over as the predominant roofing material of the Victorian age all over the country, both urban and rural – anywhere the railways reached. It was readily available in quantity and cheap. Welsh slate is grey–blue in colour and individual slates are generally just a few millimetres thick (2–3 mm). They are a lot lighter than stone slates and therefore a lot less demanding in terms of supporting roof structure.

Problems associated with Welsh slate-covered roofs include:

- cracked and/or slipped slates;
- delamination;
- dampness in related roof spaces;
- sagging roofs.

Figure 13.2 Complicated slate roof junction.

Figure 13.2 shows two buildings with adjoining slate roofs of different pitches. Although lead soakers have been installed, the junction is a difficult one and several slates have slipped.

Where slates have slipped, this is generally attributable to nail sickness. This occurs where the nails used to fix the slates in position on the battens have rusted through. Copper nails or similar would normally be used today, but the issue was not so well understood in earlier years – or if it was, matters were often skimped. A single slate could be returned to its original location and held in place by a thin metal cramp (or 'tingle') hooked over the lip of the slate, under the slate and over the batten, but this should be considered temporary. If one slate has slipped, others will be similarly afflicted and a wholesale reroofing will be required in due course. Slates may be broken where they have been walked on in gaining access to replace other slates or typically to carry out repairs to chimneys or to place TV aerials. If it is decided to reroof with a different material, its weight must be taken fully into account. While it is possible to strengthen roof structures this may have knock-on effects on the transfer of loads on to walls, lintels and foundations which may not be adequate.

Delamination occurs when frost gets between the layers within the slate. Ironically perhaps this is more likely to happen with well insulated roofs where the slate does not 'benefit' from heat lost from the building beneath. There is no way back for slates thus affected and they must be replaced.

Dampness in roof spaces under slated roofs tends to occur due to condensation on the underside of slates and the absence of sarking felt to direct dripping moisture away. This can be reduced by attention to insulation and ventilation of roofspaces; this will be important where attic spaces are to be brought into habitable use.

Sagging is unlikely to be caused by the slates themselves, but is a problem often associated with buildings of the period when slate was prevalent. Because Welsh slate was a cheap material, it was often used in conjunction with other cheap constructions. Thus roof timbers would be of such size as deemed to be adequate, and may not have been quite substantial enough. Alternatively timbers may have rotted where rain has penetrated at chimneys or through cracked or missing slates.

Clay tiles

Tiles are described by both their material and their shape. Clay tiles are common in much the same localities as clay bricks. Traditionally they were made by hand and therefore liable to fail at individual defects. With the advance of machine-made tiles in the twentieth century more systemic failures have become common. Typical failures include:

■ broken nibs;
■ cracked hips;
■ delamination.

The nibs should be holding the tiles in place by being hooked over the tiling battens. Where the nib joins the main plane of the tile is a point of weakness; when these break, tiles may slip. It is fairly easy to lift adjacent tiles and to slip in a replacement, if one is to hand. However the remaining tiles will have weathered over time, so a new tile may stand out visually.

Figure 13.3 shows tiles 'hogging' where the roof passes over a party wall and the roof structures have deflected; there are further tiles disrupted at the flashing where the roof pitch changes.

Ridge and hip tiles are normally formed and fired in batches either alongside or separately from the plain tiles they are designed to match. However, the shape of these tiles means that heat is distributed unevenly within them, building in stresses and cracks. When open to rain and frost, small fissures may be further stressed and split the tile. Replacement is the only answer; and as such weaknesses are likely to be common to a batch, it probably makes sense to consider wholesale replacement in due course.

Delamination occurs in much the same way as with slates. Traditionally, plain tiles were made with a camber to help shed the rain water, but the machine-made varieties were flatter and this has allowed moisture to be held between tiles and to freeze there, contributing to delamination. Discolouration may also be a problem as tiles weather over time. Because clay is a more absorbent material than slate, it is also prone to moss and lichen growths, which may be heavy in some areas. There are proprietary systems that claim to remove such growths and restore tiles to their original colour. Concern must be attached to any product that may 'eat into' the tile.

Not all tiles are 'plain' however. In southern Europe, pantiled roofs are common. Traditionally these are formed by alternating half-round clay units so that they overlap to direct rain water down the roof slope. These have generally stood the test of time well in that climate. They have the advantage of ready replacement in the event of failure. British versions of pantiles are generally formed to an S-shaped section and are more prone to damage and to dislocation of the roof structure and surface. As with ridge and hip tiles there is scope for fissuring and cracking. It is

Figure 13.3 A deflected and disrupted tiled roof.

much harder however to find replacements, as patterns change and cease to be manufactured; it is worth keeping 'spares' from occasional reroofings. Bridgewater tiles, a large interlocking clay tile made in Somerset, are a case in point. There was a fashion for glazed roof tiles in the 1930s and 1950s, often in vivid blues and greens, in some of the new suburbs.

Concrete tiles

The construction industry is always innovating. Concrete is a very versatile material and was introduced into roof tiling in a big way in the twentieth century. Concrete plain tiles are available as a direct replacement for plain clay tiles: same face sizes and thickness, etc. Being more durable and cheaper, they have largely supplanted clay plain tiles in all circumstances other than where the visual quality of clay was preferred. Defective clay plain-tiled roofs have largely been stripped and replaced by concrete tiles.

The 1960s and 1970s saw a radical 'advance' in roofing with the large-scale introduction of the concrete interlocking tile. These are sometimes referred to as concrete or interlocking *slates*, although this is disliked by the natural slate industry. Advantages of these units include their large unit size, enabling rapid coverage of a roof, and that they are single-lap, thus reducing the number of units required. Material failures are relatively rare; most problems relate to mechanical damage incurred through people walking on them, poor details at

Figure 13.4 Moss and lichens growing on asbestos–cement slates (also note tingle).

junctions and inadequate roof structures. The Concrete Tile Manufacturers' Association (CTMA) offers advice on the care, maintenance and repair of concrete roof (and cladding) tiles (CTMA, 2004).

A common error is to replace a clay-tiled or slated roof with concrete interlocking tiles that are heavier per unit of area. This inflicts a load on the roof that may be beyond what it can bear. Deflection in the roof will result in the joints between tiles being opened up and liable to having rain and snow enter the roof beneath. Any roof of interlocking tiles that shows disruption of its surface should be considered to have an inadequate roof structure which will need thorough inspection and evaluation. The tiles may also be inadequately fixed in position, be open to being dislodged, and therefore dangerous. The BRE has published guidance on re-covering pitched roofs (BRE, 1998a).

Asbestos–cement slates and sheets

Asbestos–cement slates were developed as an alternative to slates or clay tiles. Those used in the 1930s and 1950s were generally reddish in colour (now weathered to a dull pink) and fixed in a diamond pattern; those from the 1970s were more often grey and laid to mimic slates. They have a tendency to curl and also to absorb water; they thus attract moss and lichens (Figure 13.4; note also the 'tingle' holding a slate in place). Of course since their manufacture we have become aware of problems relating asbestos fibres to respiratory problems. Thus any problems with these slates are likely to lead toward a decision to replace the roofing wholesale, which is a significant undertaking; a considered view should be taken. Licensed asbestos contractors will need to be engaged if work is to be undertaken to such roofs.

Corrugated asbestos sheets were developed at the same time, as a material capable of roofing warehouses, barns and similar commercial/industrial structures economically. Where these sheets are cracked or broken they should be removed by a licensed contractor and replaced with alternative materials. Because all sheets must be suspect when one or more has failed it must be worth considering wholesale replacement.

Fibre–cement

This was developed in the 1980s as an alternative when asbestos came into doubt. It looks and behaves similarly to asbestos–cement. It may be difficult to distinguish between the two materials, so it may be necessary to seek out documentary evidence and to consult specialists.

Sheet metal roofs

In UK these are largely a feature of industrial 'sheds', whereas in Australasia and elsewhere they are more universally common in the residential sector too. They are commonly of lightweight steel or aluminium profiled sheets, screw-fixed and with laps over adjacent sheets. These screws typically have washers to provide some weather-proofing at the fixing. The fixings are the main source of problems, which may include:

- leaks associated with non-existent or cracked washers;
- inadequate screws;
- fixings through the troughs of corrugations;
- bimetallic corrosion between screws and sheets of different metals;
- rusting of sheets due to damage or inadequate protection such as galvanising;
- solar degradation of sheets and/or washers.

Occasionally metals such as copper or lead will be used extensively on a roof, often where the building is of some age and/or of significant architectural quality. These are discussed more fully under flat roofs, below.

Other roofing materials

Straw thatch may also be used in some areas, and then generally on old buildings (although there has been renewed interest in relation to new buildings in rural areas). East Anglia, the Midlands, south and southwest England are areas where thatch is relatively common. Thatching is a skilled craft and it is as well to consult an expert locally who will know about local details including varieties and sources of straw.

Turf is a traditional roofing material in 'celtic' Britain: the highlands and islands of Scotland and Ireland. Again interest is growing, through the development of 'green roof' technology.

Roof forms and related details

Roof shapes and forms are very much determined by the plan-shape of the building beneath. At the risk of over-simplification, simple rectangular buildings tend to lend themselves to simple dual-pitched roofs with simple rainwater disposal arrangements. Plan depths greater than 10 m can become more problematic, in that ridge heights tend to become substantial unless particularly shallow roof pitches are used, or more complex and expensive roof structures are required.

A dual-pitched roof with gable walls at both ends is the most straightforward, although a simple hipped roof may be preferred. When building plans are L-shaped or more complex, or step up or down, or roofs are to have laylights or dormer windows within them, roof forms and details become more complicated, and more prone to failures. There is more scope for failures at junctions than within the main run of roof.

Ridge and hip tiles have been discussed above; problems with them tend to relate to cracks in the components. Rainwater tends to be shed by these components, by contrast with valleys and eaves to which water is led by gravity.

Valleys

Lead has long been the traditional material for valleys. Problems tend to relate to its wear or slump or poor fitting, or to its lack of support, as described in relation to chimney flashings and aprons. If roof coverings are being replaced it is sensible to consider replacing valleys also, as they will never be more accessible than at that time. Repairs to valleys can be difficult as access is restricted and working conditions awkward. If valleys are of a material other than lead and have failed it is worth considering upgrading the specification for their replacement. Lead is generally favoured because it is durable and malleable. It is however prone to creep so it must be properly fixed. The Lead Sheet Association (LSA) provides excellent advice (LSA, 2007).

Valley gutters behind parapet walls are particularly problematic. Their falls are generally shallow and they are difficult to access, often requiring to be walked in. There should be substantial upstands beneath the roof finishes that discharge into them, to prevent snow from entering should it build up in the gutter. Such valleys are very prone to filling up with leaves and other debris and should be cleaned out at least twice a year. Any penetration will lead to water ingress with potentially dangerous consequences if it should enter electrical systems or cause rotting of roof timbers.

Parapet walls are more exposed than the walls below them, both because they are higher and because they are exposed on both sides. Thus the same brick is likely to deteriorate sooner; spalling is more common. Parapet walls are wetter than the walls of the building or of chimneys – they are unheated. Water standing in a valley gutter at the base of a parapet wall will also provide a rich source of moisture to be taken up into the masonry should there be such possibility afforded. They should be inspected rigorously and repaired carefully (BRE, 1998b); a piece of masonry falling from a parapet will be dangerous.

Figure 13.5 shows dampness in an external wall almost certainly stemming from problems in the parapet gutter behind. One of the parapet walls in Figure 13.6 has moved out of plumb (or possibly both have); further investigation is required.

Water from valley gutters should fall to adequate and appropriately placed outlets, which also need to be kept clear. These points are particularly liable to damage and leaks here will involve more water than at higher points so they must receive special attention.

Remember also that lead is poisonous so should not be handled with bare hands. Do not eat your lunchtime sandwiches without washing your hands thoroughly.

Eaves

Much the same considerations apply as for valleys. Rainwater goods are considered more fully below. Suffice it to say here that this coming together of roof surface, roof structure,

Figure 13.5 Damp brickwork emanating from parapet gutter problems.

rainwater goods, external wall and window heads is complicated and the scope for water ingress if it is not got right is considerable. All the water that falls on a roof is concentrated at the eaves.

Problems likely to be manifest at the eaves include:

- missing or misaligned roof tiles or slates;
- missing or displaced or torn or unsupported sarking felt;
- misaligned or incorrectly sized gutter;
- blocked gutter or rainwater outlet or hopper or downpipe;
- broken gutter;
- rotten fascia and/or soffite and/or boxing out;
- damp brickwork or stonework;
- dangerous bulging and/or loose masonry.

Figure 13.7 shows slates slipped at the eaves (and vegetation growing on the roof and in the gutter). Figure 13.8 is a good example of a response to a difficulty – rainwater that was previously overshooting the gutter and causing problems is now restrained.

Because of the complexity of arrangements at eaves, investigations need to be thorough, considering all possibilities of contributory causes and consequences. Because of difficulties of observation and access it is also worth considering doing more 'while you're there'.

Figure 13.6 A parapet wall problem? Further investigation is required.

Figure 13.7 Slates slipping at the eaves (and vegetation growing in the gutter).

Figure 13.8 An innovative solution to a problem.

Dormer windows and other penetrations

Dormer windows are very much like chimneys in relation to their implications for roofs. The penetrating structure needs to be properly flashed on all sides; these intersections are points of weakness and possible ingress for water to the roof structure. They should therefore be checked regularly and repair/replace decisions taken accordingly.

Dormer windows are more exposed than windows in the main structure of the building; they are therefore likely to show problems sooner. The surrounding and supporting structure of the dormer is also likely to be insubstantial and liable to deterioration, decay and distortion. The window too may be of inferior specification and construction compared with those in the main structure. Roof coverings are also likely to be more problematic, as the smaller areas and complicated junctions tend to mean more cutting (with consequential scope for error) and more joints. Dormer windows need to be well designed, well detailed and well constructed; often they are not. What must be kept in mind here is:

▪ Where should rainwater be going?
▪ Where might it go?
▪ Where is it going?

Laylights give similar problems. There is a sense in which the light is like a large slate or tile but of glass, notionally impervious but with joints all round that render it vulnerable to water

Figure 13.9 Flashing or fillet: render details (note also delamination of clay tiles).

ingress anywhere around it. Flashings may be missing or broken. Supporting structure may be overstressed and/or rotting. Glass may be cracked. Condensation may create problems.

All penetrations of a roof introduce vulnerabilities. Soil vent pipes (SVPs) should be properly sleeved with an upstand skirting all round and a cover fillet dressed over so that water cannot enter. It is better to avoid all such penetrations if possible.

Steps – flashings

Where roof levels step down from one part of a building to another, for instance between terraced houses, or between a two-storey and single- or three-storey structure, flashings must be provided to the vertical structure. At the higher level the roof will normally oversail the wall structure at that point, and at the lower level it will abut the wall. The wall is vulnerable to the ingress of water.

The wall may be of a cavity construction, in which case there should be a combination of cavity trays at intervals following the pitch of the roof and stepped flashings dressed into the wall and on to or under the roof covering, depending upon the detail. If there are problems of water ingress at such an intersection it may be necessary to remove some of the structure in order to gain a good view of what has actually been built. If the wall is of solid masonry construction, it is more difficult to avoid problems. Problems often occur when adding an extension to a solid-walled building; it is best to ensure that water falls away from the existing structure.

Figure 13.9 illustrates a common situation. The render on the right-hand house has been applied down to the tiles that roof the ground floor bay window; the house to the left has had a lead flashing installed. Whether this is as a result of dampness problems is not known. (It can also be seen that the clay tiles on the right-hand house are badly delaminating on their exposed surface.)

Gables

Occasionally problems occur with gable walls. One such problem is where, seized by economy, the wall thickness is reduced above the level of the inhabited space. This can give problems related to slenderness, and if the roof suffers movement this may be transmitted to the gable below. Gable walls should be constructed as an integral part of the wall structure and roofs should be tied in to resist them lifting in high winds. Gable walls can be sucked out by high negative wind pressures and vortices, particularly where winds are funnelled around and between buildings. If gaps are observed in roof finishes at gable ends, these may be indicators that this possibility should be investigated.

All outward signs of problems should be followed up internally whenever possible. This will mean taking good notes as memory may be fallible, and adding to them as the problem becomes clearer or more complicated and/or additional issues are observed.

Attics and roofspaces

Sometimes spaces within roofs have been used very productively to create valuable accommodation within a building; in others the space is just a void, maybe even inaccessible for inspection. It is important that they be inspected where possible as they can be rich sources of evidence and indicators of problems in and on the roofs within which they are located, and where there may be no outward sign.

Roof structures are as varied as the buildings of which they are a part, and the use of roofspaces often will have entailed the modification of structure in ways that may not have been anticipated.

Pitched roof structures

The most common structural forms of pitched roof comprise:

- rafters and purlins;
- trusses;
- trussed rafters.

The first two categories here may be described as 'cut roofs', in that the timbers were traditionally assembled and cut to size on site. Trussed rafters were a 'modern' development of the 1960s, prefabricated in a factory and delivered to site ready made, needing only to be lifted into place and fixed in position. Steel roofs are also a possibility, more often in larger, industrial, commercial or leisure buildings. There are other possibilities too, such as concrete portal frames for barns and 'glulam' structures, of which more will be said later.

Rafters run from ridge to eaves and are of such sizes and spacings as to support the roof covering and its loads, such as people working on it and snow. Sometimes it is more practical and/or economical, depending upon the spans and available timbers, to introduce purlins intermediately to reduce the spans and thereby the sizes of rafters required. Normally, but not always, these timbers will be hidden behind ceilings, traditionally of lath and plaster or more recently plasterboard. Sometimes the timbers are exposed. Sometimes insulation material has been introduced between the rafters.

An advantage of this form of roof is the relatively free access to all parts of the roofspace, although this may have limited headroom and there may be difficulty in seeing into the eaves. The roof timbers and the underside of roof coverings can therefore often be inspected fairly readily, especially if the roofspace is inhabited. If the space is not in use, it may be dark and inhabited by spiders, pigeons, mice and rats, so care is needed. It is not good to put your foot through the ceiling; indeed it can be very painful as the shattered laths dig into your leg. **Do not inspect on your own.**

Trusses, because they usually have larger timbers and are at wider intervals, can also provide quite good access, dependent upon the placing of struts and other intermediate members. Trussed rafters however are often a 'sea' of small timbers, usually in a low pitched structure and therefore virtually inaccessible for anything other than storage of junk and access to a cold water tank.

Problems that may be observed in a roofspace include:

▒ damaged timbers – rot, insect damage, gnawing;
▒ timbers adapted by sawing, notching, bolting, etc.;
▒ timbers missing;
▒ sagging timbers;
▒ water ingress;
▒ dampness/staining;
▒ torn or incomplete sarking felt;
▒ missing or misplaced tiles or slates;
▒ incomplete insulation;
▒ chimney problems;
▒ fire-related issues;
▒ electrical problems.

The first of the matters listed above is substantial; the possible consequences of such damage are very serious and potentially dangerous and there is much that could be written. Not only has much already been written but there are also many specialist firms that will be happy to inspect and provide recommendations for timber treatment.

I shall endeavour to summarise matters here. The key issue is that not only should decayed material be removed but that sources of problems should be dealt with so that problems do not recur.

Rot

Sometimes rot is very readily apparent; in other situations it may be lurking behind a surface which may be giving some indications of the problem beneath. *Wet rots* generally occur in exceptionally wet locations. The timber looks wet and it is wet; and the sources of wetness may be fairly readily apparent, for instance rain entering the roofspace through cracked or missing slates or tiles, directed by tears in the sarking felt or water from blocked and overflowing gutters. If caught early and the source(s) of the wetness is dealt with, it may be possible to dry out the timber satisfactorily; if not, the affected timber should be cut out and removed. In that case attention will need to be given to how the replacement timber is to be introduced and jointed in so that structural performance is not impaired. Temporary support may be required. This needs to be fully thought through so that loads are not transferred to members that are inadequate for the purpose.

Dry rots are also caused by wetness and are often found in areas where dampness is accompanied by still air conditions. They are called dry rots because affected wood often looks dried out and desiccated, with cuboid looking cracks on the surface of hollowed out, powdery and crumbling timber. When affected linings or panels are removed, grey–white threads or hyphae may be seen, perhaps accompanied by fruiting bodies and spores of pale grey, slightly yellow and rust-red. This may look rather dramatic and alarming and it is appropriate to be alarmed. Affected timber must be completely removed (extending to at least 300–500 mm beyond what is seen) and burnt, on site if possible. The fungus is a virulent one and can travel extensively. Surrounding brick and stonework will also be affected and must be effectively sterilised. All signs of the outbreak must be eradicated and all causes eliminated. Often this will mean changing the environment in which the problem has occurred. This may entail the provision or improvement of ventilation and the removal of sources of water such as by redesigning rainwater systems. Inaccessible areas are particularly vulnerable.

Other enemies of timber

Insects can infest and seriously affect timber. Their presence is normally apparent from frass on floors around and below flight holes of beetles in affected timber. The damage may not be of itself structural, although it may be, but it is certainly disturbing visually. Outbreaks must be treated promptly. In most modern properties roofing timbers will have been pre-treated against beetle attack, but sometimes cut ends are not treated as they should be and non-structural timber, such as linings and panels, may be untreated. Infestations may be 'imported' in affected items of furniture. A useful review of insect enemies can be found at www.rentokil.co.uk/pest-guides.

The structural integrity of roof timbers may also be impaired by the actions of animals such as rats and mice and by humans. Bats and birds may also be found, but while their presence maybe inconvenient (bats are protected by UK legislation), they tend not to be destructive of structure. Rodents will gnaw at timber and may cause serious damage; human beings may also cause damage. People, particularly do-it-yourself enthusiasts but also including unqualified 'builders', sometimes saw, drill and chisel out pieces of timber to provide routes for services and access. Cuts and notches in timbers under tension are especially vulnerable and should be viewed with suspicion. Joints, including those provided by gangnail plates and similar connectors, bolts and nails should also be inspected. Nails are liable to cause splitting in timber, and connectors may succumb to rust. Occasionally, timbers may have been completely removed without consideration to their contribution to the roof structure. Roofs of trussed rafter construction may not have the requisite binders or bracing that should be provided to give stabilty to the structure – this should be checked against advice from the Timber Research and Development Association (TRADA, 2007).

Rafters and purlins may be seen to be seriously deflected. In a particularly old building, before rectangular sawn timber became the norm, such timbers may have been rough hewn and of a shape and section closer to that which it had in the tree from which it came. Although such timbers may have distorted over time they may now have achieved an equilibrium in which they are satisfactorily bearing the loads to which they are submitted. Other timbers may have been undersized from the outset, and others may be bearing loads now beyond those for which they were intended. An example of this latter would be where original roof coverings which may have failed have been replaced with heavier ones. Another possibility is that the deflected timber has become defective due to mechanical damage, splits or rot.

Other issues

Any evidence of incompleteness of the weather-resisting envelope or of actual water ingress, whether present or past, should be investigated and any current deficiencies rectified. The presence, thickness and completeness of thermal insulation materials should also be checked. Most buildings will benefit from having more insulation than they have; standards and energy costs are constantly increasing, so it can be a very cost-effective improvement. There can be problems though. Insulation at the 'floor' level will result in a cold roofspace (unless it is heated) with consequential possibilities of freezing of water pipes; whereas insulation at the roofslope will mean the roofspace becomes heated by 'waste' heat from the rooms below. If insulation material is pressed down into the eaves it may inhibit the ventilation required to avoid condensation in the roofspace (BRE, 1982a) – this should be inspected with care as fibres from the insulation can cause irritation. Completeness of the sarking felt should also be checked (BRE, 1982b, 1982c).

Cold water storage tanks are often located in roofspaces. They could leak, especially where pipes enter or leave, although problems that look like leaking could stem from condensation. Tanks should be insulated, and should not have insulation beneath them – they need some heat from below to prevent them from freezing and bursting. Water is heavy, so the weight of a water tank needs to be distributed across supporting structure. Normally the tank will be supported on a board or boards on bearers and cross-bearers placed across a number of joists to spread the load (BRE, 1984a). Inadequate support is likely to show through excessive deflection of the supporting members.

Roofspaces can also be sources of problems related to fire protection. If, for instance, the building has acquired another storey by virtue of conversion of the attic to useable accommodation, more stringent provision, including means of escape, will be necessary. They can be a rich source of services-related problems too. Often service pipes and wires pass through the roofspace where they are vulnerable to damage and/or neglect – a visual check may provide impetus to a fuller evaluation.

Flat roofs

Flat roofs, or more accurately low-pitched roofs, have been around a long time. Lead has long been used as a flat roof covering. Felt and asphalt are relative newcomers and have contributed largely to bringing the flat roof into disrepute. Problems with flat roofs tend to relate to materials used and to the comparative lack of fall.

Lead

By and large, lead is a durable and long-lasting material. On flat roofs it is generally laid in overlapping sheets, with joints designed to allow for thermal movement, which can be quite appreciable. Clearly problems can arise if these joints have not been correctly executed. Lead-work, including its repair, should be undertaken by specialists; they should know what they are doing.

Problems that may arise with lead roofs include:

■ fatigue due to excessive expansion/contraction, leading to cracking;
■ damage from chemical reaction with moss or lichens, cedarwood, cement or concrete;
■ expense and difficulties of access for repair work;
■ high material resale value makes it attractive for thieves.

Felt

This is a cheap material. Generally it is applied as two layers, or, better, three of bituminous felt, to reduce the risk of puncturing and to stagger joints. The joints and fixings are points of

Figure 13.10 A felted flat roof to hospital building.

weakness against water ingress. Poor falls also contribute. Although felt roof finishes should be laid at a minimum fall of 1 in 60, in practice this may not have been achieved, and subsequent movements of the supporting structure may mean that in fact there is now a backfall, with resultant ponding. Standing water will threaten to enter through any crack that appears in the felt, which is very likely to occur. Felt is not a suitable roof covering for anything other than a shed roof; plans should be made for its replacement with something more durable, at which time the opportunity should be taken to upgrade insulation too. The BRE offers advice on remedial work (BRE, 1983a, 1983b, 1998c).

Figure 13.10 shows a 'lake' forming on a felted flat roof; falls have proven inadequate to ensure that the rainwater reaches the outlets provided and any puncture of the felt will result in serious water ingress. Figure 13.11 illustrates an unusual example; here the felt roof covering has been continued down the vertical face between two levels of roof and must be problematic.

Asphalt

Asphalt roofs are more durable than felt roofs though they share many similar features. Their longevity is contingent upon effective detailing and provision of upstands and movement joints, which are often problematic. The material itself is also prone to problems related to protection from solar degradation over time. Blisters, cracking and crazing are common problems. Asphalt roofs should be inspected at least annually. Defective or problematic areas should be cut out and made good with new asphalt. Areas for particular concern include:

- penetrations such as rooflights, smoke extracts and vent pipes;
- roof structures such as lift motor rooms and other services installations;
- upstands of any kind;
- edge details.

Figure 13.11 An unusual detail to a felt roofed building – a school.

BRE offers advice on assessing felt and asphalt roofs (BRE, 1998c) and their repair (BRE, 1998d). Consideration should be given to the long-term prognosis for the roof.

Insulation

The provision of thermal insulation in flat roof structures has often been problematic. The relative virtues of 'cold roofs', 'warm roofs' and 'inverted roofs' are not discussed here; those are design issues. Suffice it to say here that whatever has been provided has to be made to work as effectively as possible. Where insulation has been provided at or near the roof surface it is likely to have been damaged; this should be checked as areas of reduced insulation will give rise to cold spots and possible condensation. Most insulation as provided in any building more than just a few years old will be woefully below today's standards. All roofs that are requiring attention should be considered for upgrading of their thermal insulation; virtually all roofs would benefit. The roof is the most cost-effective place for improving thermal insulation.

In roofs where the insulation is relatively inaccessible, 'sandwiched' between other materials, there may be cold spots where there are gaps; this may be revealed by a survey using thermal imaging technology. Remedial work may be difficult and expensive. Again, it will be worthwhile to consider upgrading the insulation by the addition of further layers above or below the existing roof structure. Assessment should be made of the anticipated performance of the new arrangement including condensation risks, the logistics of carrying out the work, and of course the costs in order to determine the cost–benefit equation.

Protection

Workers on flat roofs, including surveyors – indeed all who have access to the roof – need to be protected from falling from it. Perhaps the most common provision is a solid parapet wall; other common arrangements are metal or stonework balustrades. All these can be problematic. The structure has to be capable of resisting lateral loads from falling or pushing and this stresses its junction with the roof and building below. Damp proof courses at this point provide a potential slip-plane; the absence of a damp proof course will permit water penetration. Loose copings can also be a problem.

Fixings of metal balustradings may penetrate roof surface coverings; they may rust, expand and 'blow' surfaces. The vulnerability of these areas is exacerbated if accompanied by adjacent gutters in the flat roof surface or if rainwater is otherwise directed that way by virtue of its falls. Effective upstands are required to keep fixings out of wet areas. If the roof has not been well designed from such points of view it will be expensive to remedy, but the safety of people on the roof or passing by below must be the paramount consideration.

Roofs of 'on-roof structures' such as lift motor rooms, and access to them, also need to be considered.

Proprietary systems are available that allow for workers on flat roofs to be hooked on by cables to plates fixed to the roof surface.

The Flat Roof Alliance has prepared guidance (Flat Roof Alliance, 2007) regarding safety in flat roof refurbishment.

Green roofs

These have been receiving increased interest as a mechanism for not only improving thermal insulation in a particularly 'green' way but also attenuating the discharge of rainwater. Clearly, designs must take account of issues related to loading and waterproofing, amongst other matters. Tracking down the causes of and rectifying defects may be more problematic than with more traditional constructions; time will tell.

Rainwater disposal

Rainwater must not be allowed to settle on or near buildings; it is potentially destructive. Traditionally, pitched roofs have so directed rainwater down the slope of the roof and away from the base of the building; and flat roofs have employed gargoyles to concentrate and shoot water away from the building. More recently, gutters at the eaves have become the norm.

As already noted, where there are parapets provision must be made for rainwater to be collected and directed away, either through outlets through the roof structure or by chutes through the parapet wall to an external hopper or gutter. These need to be well designed and constructed; they should be inspected carefully, frequently and regularly to prevent blockages.

The rainwater disposal system can be said to be defective if it has failed to take some or all of the water away from the building, or is likely to fail so to do. Contributory factors could include:

■ cracked gutter, hopper or downpipe;
■ poor location of gutter in relation to where rain is discharged from the roof;
■ inadequate design, e.g. undersized;
■ missing or fallen sections;
■ blockage;
■ failures of fixings;
■ underground drainage problems.

Figures 13.12–13.17 illustrate a range of rainwater disposal issues. Figures 13.12 and 13.13 show complexities of collecting rainwater from more than one roof into a shared downpipe.

Figure 13.12   A hopper receiving rainwater from two roofs.

Figure 13.13   Rainwater from one roof discharged over another.

Figure 13.14　A complicated gutter detail.

Figure 13.14 is of a purpose-made pre-formed pitchmastic gutter; the 'hidden gutter' detail at the gable parapet is a difficult design and difficult to execute successfully. Rainwater can be seen to be saturating the brickwork of the adjoining gable. Figures 13.15 and 13.16 show common problems – a dislocated spigot joint and a missing rainwater shoe – while Figure 13.17 illustrates clearly the effect of what one would think would be an only too obvious shortcoming.

Cracked components will often be, though not always, readily apparent; they may be masked by accumulated dirt. Indeed, sometimes cleaning and clearing of material may open up problems, enabling cracks to leak that had been to an extent 'sealed' by dirt that had filled the crack.

Blockages may be revealed through cracks or joints, especially in downpipes. Rainwater flow may be 'backed up' by a blockage as far as such a crack or joint where it then discharges. Clearly blockages must be removed, which is not always easy or straightforward. Easy access is not always readily available. Tennis balls, twigs and leaves are common villains. Whenever a blockage is removed, steps should be taken to prevent a recurrence.

Gutters are often poorly positioned. When rainwater is discharged from the roof slope it falls on a trajectory somewhere between the pitch of the roof and vertical; it may even dribble back a little from the lip of the tile or slate or roof edge depending upon the detail. The trajectory will depend upon pitch, the amount of rainfall and rate of discharge. Some will discharge from the sarking felt into the gutter. It is difficult to assess the optimum position, also taking into account

Figure 13.15   A dislocated spigot to a cast-iron downpipe junction.

the fall of the gutter along its length. The greater the interval between downpipes, the longer the gutter, the greater the fall (or the more shallow!).

Discharge from valley gutters can be particularly problematic, especially from large roof areas with greatly concentrated volumes and flows. Overflowing gutters are evidence of under-design. With climate change we should assume that any overflowing will not be an exceptional occurence but an increasingly likely recurrent event, demanding a review and redesign of the rainwater disposal system.

Lead was commonly used for gutters, hopperheads and downpipes on buildings that are now regarded as historic; they often contribute strongly to their quality, including decorative detail, dates, etc. Clearly such materials deserve to be maintained. Victorian and Edwardian buildings often feature cast iron rainwater goods; generally the downpipes are round in section, and gutters may be halfround or 'ogee' in section. This latter, though decorative, is often problematic in that they are fixed to the fascia (if there is one; otherwise to the structure, maybe a wallplate) through the back face of the gutter. This tends to rust; and the fascia to rot. It is often sensible therefore to think in terms of replacement despite the visual degrading that may ensue. There are specialist firms that can provide new aluminium gutters to match the ogee profile.

The twentieth century saw material developments. Austerity brought forth asbestos–cement rainwater systems. This may seem a strange material to use bearing in mind its water-absorptive

Figure 13.16   Missing a shoe.

capability and unsightly appearance but its use has been quite extensive. While official advice has been that the material is safe if undisturbed, any failure such as cracking may raise doubts and it will make sense to consider replacement. Indeed because asbestos removal should be carried out by licensed firms, with material removed to licensed facilities for disposal, a concerted programme will make sense.

A particularly idiosyncratic product of the immediate post-war period was the 'Finlock' gutter (Figure 13.18). This was a proprietary system combining eaves and gutter in interlocking concrete units. Perhaps unsurprisingly these have not proved waterproof; their removal would involve a lot of work and proprietary systems have been developed for lining them. They are quite common on council housing of the period.

PVCu rainwater systems are now the prevalent provision. They are versatile and cheap. Problems often stem from the assumption that anyone can fit them, which is often done poorly. Early systems have tended to bleach and become brittle under the influence of ultraviolet rays present in sunlight. Thermal expansion and contraction of the material also bring problems; in Britain's changeable weather there can be quite a lot of audible creaking as the sun appears from and disappears behind clouds.

The BRE provides good advice on the installation, repair and replacement of rainwater goods (BRE, 1984b, 1997).

Figure 13.17 Missing a downpipe.

Figure 13.18 A Finlock gutter – combined eaves/gutter units, with asbestos–cement outlet.

Ladders should never be leant against rainwater goods; they do not give reliable support and they may distort or break. Nor should work be carried out to rainwater goods from a ladder – the temptation to stretch out and over-topple is too great. Access towers or scaffolding or a 'cherry-picker' should be used. There will be expense attached to this, so it is again worth considering undertaking a quantity of work at the same time.

Problems can also occur where rainwater systems discharge into the below-ground drains. There may be design deficiencies at the junction, problems with the gully or rainwater shoe (either of which may be blocked), or from the absence of a gully and difficulty of access at a vulnerable place. The drain may be blocked and backing up. If the drainage is a single-pipe system (that is to say without surface water drainage separated from foul) there is substantial scope for blockage, not only from silting but also from grease and other material from kitchens, bathrooms and toilets, and for these to back up. A thorough cleansing may be required.

Summary

This chapter represents the completion of our examination of buildings' exteriors. Our concerns have been largely related to issues of wind- and weathertightness. We now move inside the building, where the issues relate more to matters of safety and comfort.

References

Building Research Establishment (1982a) *Pitched Roofs: Thermal Insulation near the Eaves: Defect Action Sheet 4*. BRE, Watford.

Building Research Establishment (1982b) *Pitched Roofs: Sarking Felt Underlay – Drainage from Roof: Defect Action Sheet 9*. BRE, Watford.

Building Research Establishment (1982c) *Pitched Roofs: Sarking Felt Underlay – Watertightness: Defect Action Sheet 10*. BRE, Watford.

Building Research Establishment (1983a) *Flat Roofs: Built-Up Bitumen Felt – Remedying Rain Penetration: Defect Action Sheet 33*. BRE, Watford.

Building Research Establishment (1983b) *Flat Roofs: Built-Up Bitumen Felt – Remedying Rain Penetration at Abutments and Upstands: Defect Action Sheet 34*. BRE, Watford.

Building Research Establishment (1984a) *Trussed Rafter Roofs: Tank Supports – Installation: Defect Action Sheet 44*. BRE, Watford.

Building Research Establishment (1984b) *Roofs: Eaves Gutters and Downpipe Installation: Defect Action Sheet 56*. BRE, Watford.

Building Research Establishment (1989) *Domestic Chimneys: Rebuilding or Lining Existing Chimneys: Defect Action Sheet 138*. BRE, Watford.

Building Research Establishment (1990a) *Surveying Masonry Chimneys for Rebuilding or Repair: Good Building Guide 2*. BRE, Watford.

Building Research Establishment (1990b) *Repairing or Rebuilding Masonry Chimneys: Good Building Guide 4*. BRE, Watford.

Building Research Establishment (1997) *Repairing and Replacing Rainwater Goods: Good Repair Guide 9*. BRE, Watford.

Building Research Establishment (1998a) *Re-covering Pitched Roofs: Good Repair Guide 14*. BRE, Watford.

Building Research Establishment (1998b): *Repairing Chimneys and Parapets: Good Repair Guide 15*. BRE, Watford.

Building Research Establishment (1998c) *Flat Roofs: Assessing Bitumen Felt and Mastic Asphalt Roofs for Repair: Good Repair Guide 16: Part 1*. BRE, Watford.

Building Research Establishment (1998d) *Flat Roofs: Making Repairs to Bitumen Felt and Mastic Asphalt Roofs: Good Repair Guide 16: Part 2*. BRE, Watford.

Concrete Tile Manufacturers' Association (2004) *Care, Maintenance and Repair of Concrete Roof and Cladding Tiles*. CTMA, Leicester.

Flat Roof Alliance (2007) *Safety in Flat Roof Refurbishment: Information Sheet No. 30*. FRA, Haywards Heath, Sussex.

Lead Sheet Association (2007) *Rolled Lead Sheet – The Complete Manual*. LSA, Tonbridge, Kent.

Timber Research and Development Association (2007) *Trussed Rafters: Wood Information Sheet 1/29*. TRADA, High Wycombe.

Helpful website

www.rentokil.co.uk/pest-guides. Accessed 24 February 2009.

14 Defect recognition and rectification
Floors, stairs and internal walls

The preceding four chapters have considered general principals of building inspection, recording and rectification of common defects. This chapter looks at the insides of buildings. There are generally fewer problems internally and they are generally, but not always, less deleterious to the building as a structure. They can however be very serious matters of health and safety – for instance, stairways are one of the most dangerous areas of a building.

Ground floors

Traditionally ground floors were of earth, then stone slabs – problems with these were straight-forward; stones could be relaid if the surface became unsatisfactory. Subsequently timber became common, generally supported on brick sleeper walls. The solum below the timber needs to be ventilated to guard against rot. Usually this is provided by metal vents or air bricks at the perimeter, with the sleeper walls of honeycomb construction to allow good through ventilation. The exterior vents are common sources of problems. They can be broken or blocked by vegetation or high ground levels. The Society for the Preservation of Ancient Buildings (SPAB) has issued advice on the care and repair of old floors (SPAB, 1999).

Timber joists are liable to rot, due to lack of ventilation and/or at their ends by embedment into damp walls. Rotted timbers need to be cut out and replaced; new timbers could be hung on galvanised steel joist hangers (if there is room for them to be accommodated) or they will need some alternative means of support.

Figure 14.1 illustrates a situation where subfloor ventilation was insufficient. Surrounding ground levels had been increased by the addition of a concrete path, and although 'slots' had been left at the ventilation grilles these were inadequate; additional ventilation was provided by the introduction of terracotta airbricks above the level of the concrete (but still in the floor zone!).

Where timber ground floors have rotted, and especially if high surrounding ground levels may be a factor, it is worth considering replacement with a concrete floor. This could be provided either as a ground-bearing slab on blinded hardcore built up in compacted layers, or as a suspended concrete floor. The Building Research Establishment (BRE) has produced guidance on the assessment and repair of timber floors (BRE, 1997b, 1997c) and for the repair and replacement of ground floors (BRE, 1998a). There are many proprietary suspended floor systems – the selection criteria will need to include for ease of getting the units into place in the confined structure of the existing building.

Existing concrete floors may have problems. Ground-bearing slabs may suffer problems such as subsidence, unevenness and hollowness. The floor may be tipping toward one side or a corner

Figure 14.1 Inadequate subfloor ventilation, and an additional airbrick.

from which ground may have subsided. The adjoining wall(s) may or may not have subsided too, depending upon the adequacy of the foundations. Surfaces may be uneven due to poor quality of the initial construction, and hollowness (detectable by a ringing sound when the floor is struck) may be evidence of the opening up of a mineshaft or well below the floor. In some mining areas, shales (an unwanted by-product from mining activity) were used to support concrete ground-bearing slabs; in some places there have been chemical reactions which have necessitated their removal (Department of Communities and Local Government (DCLG), 2008).

Floor finishes

A timber suspended floor will normally have a timber boarded finish to it. In older buildings these may be butt-jointed. Over time and with timber shrinkage these joints can open up and provide a rich source of draughts. Open joints should be caulked allowing for future expansion and contraction as moisture content changes. Tongue-and-groove flooring overcomes that problem, although again it is important that moisture movement be allowed for by not boarding right up to edges, and masking these with skirtings of appropriate thickness. Because there are fewer gaps in a tongue-and-groove floor, there may be more problems if ventilation is inadequate.

Sometimes, chipboard or other panelboards are used to finish a ground floor. In such situations, chipboard should be moisture resistant and all cut edges should be treated – normal grades are very vulnerable to moisture attack and decay. These are cheap materials not well suited to the situation. Access to underfloor areas where services may run is also often problematic unless it has been specifically planned for. Large sheets may be economical to lay; they are a nuisance when access is required.

Concrete floors will generally have a screed of some kind to provide a flat and level upper surface. Often these screeds crack through uneven drying or inability to accommodate movement or under unexpected loads. This can be unsightly where the screed is the finished surface. Materials are available to fill cracks and/or to provide a new smooth finish; guarantees should be sought from contractors (BRE, 1997a; Construction Industry Research and Information Association (CIRIA), 1998a, 1998b).

Some early thermoplastic floor tiles have not stood the test of time, succumbing to brittleness and cracking. Usually the designs went out of fashion and out of stock many years ago and wholesale replacement must be entertained. Vinyl sheeting has generally provided a good surface, though some architects have become reacquainted with linoleum and have come to appreciate its 'environmental' credentials.

In some parts of the world, ceramic floor finishes are *de rigeur*. Good standards of finish are contingent upon a skilled labourforce. Many Victorian houses in UK have very fine decorative tiled floors, especially in their front porches and entrance halls and these have become quite prized features. It is certainly unlikely that such decorative assets will be features of speculative housing estates in the foreseeable future, so it is well worth trying to maintain such characteristics:

■ Cracks in tiles or grout should be attended to promptly.
■ Soapless detergents should be used – soap leaves a film on which mould may grow.
■ Do not use steel wool or abrasive cleaning materials.
■ Consult experts as different materials require different cleaning regimes.

Intermediate floors

These are generally less problematic than ground floors; they are not in contact with the ground and all its problems. They do however share some of the characteristics and related issues. In houses, intermediate floors will normally be of timber, with timber floorboards on timber joists. Joist ends may rot in the same way as for ground floor joists if they are embedded into external walls. If this should be the case, joists could be replaced with support being gained from galvanised joist hangers built into the supporting wall. This is not easily done, but it is possible; care must be taken that the hangers are properly supported in and by the wall.

Timber joists may be inadequately sized for current loadings. It is possible to strengthen timbers – a structural engineer should be consulted. Herringbone strutting or solid bridging between joists may help provide greater rigidity of the floor structure.

Where a house is converted into flats or maisonettes, timber floors will be required to give improved fire and sound resistance. Often this can be provided by the addition of layers of plasterboard to the ceiling, but this may be difficult where ceilings are particularly decorative. It is important to discuss such issues with building control personnel so that due consideration is given to the architectural qualities and challenges of the building. It may be that a holistic consideration of fire resistance and means of escape will be appropriate and that both appearance and utility can be respected without endangering occupants. In large old buildings, such as country house hotels, the arrangements of floors and rooms can often be quite complicated. Bearing in mind that occupants of those rooms are likely to unfamiliar with the layout and means of escape in the event of a fire, I do not recommend relaxation of requirements there, especially as their architectural quality is often already degraded.

In purpose-built blocks of flats and in commercial properties, such as purpose-built hotels and offices, concrete floors are the norm. Normally these will provide good levels of fire and sound resistance in accord with the requirements of their day. Standards have not changed much in these areas over the years. Suspended ceilings may require attention and there may be occasional situations where integrity of fire resistance of a floor may be impaired by its penetration by services, soil pipes etc.; these latter are dealt with in the next chapter. Problems with suspended ceilings may relate to damaged or missing tiles (which may require replacement over a larger area than hoped for if spares are not available or if new tiles are so much fresher looking than the existing as to stand out unacceptably) or to the suspension system. Issues there are likely to focus around visible deterioration of the hangers and rails. Changing fashions may be a factor too.

Stairs

These can be quite complicated and rich sources of problems. That they are dangerous areas has already been remarked upon. A particular danger is where stairs have winders within them. If new stairs are being provided, every attempt should be made to redesign so as to eliminate the need for winders. If they are unavoidable, they are better placed at the bottom of stairs than at the top where the consequences of misplacing one's step are that much more serious. BRE has published advice in relation to the safety of stairways (BRE, 1984).

Stair treads, the part of the stair on which the foot is placed, should be of adequate size and in good condition. Treads are liable to wear, especially where the stair is in constant use. Treads should therefore be checked for wear periodically. They are generally more worn at the nosing where the foot tends to 'roll over' on the edge. Nosings can be provided or replaced in hardwood; care then needs to be taken that the whole of the tread is adequately supported in the staircase: this may require the nosing to be tongued into the body of the tread to support their edges.

Care is needed where carpet has been fitted. If the tread beneath is uneven this could be dangerous.

Sometimes timber stairs develop a squeak. This is normally rectified by tightening up screws fixing treads and risers and/or by rewedging their housings into the strings. It may also be beneficial to provide a fixing between the string and the wall if a gap has opened up, thus providing greater rigidity and resistance to movement.

In buildings with concrete intermediate floors, concrete stairs are the norm. Problems with concrete stairs tend to be related to damage. As with timber stairs, nosings are vulnerable; proprietary edgings can be provided and these should be selected with finishing materials in mind for they should be fitted into the back edge of the nosing. Occasionally there may be problems with spalling concrete. These may relate to poor placement of reinforcement so it is important to carry out an investigation promptly and to respond accordingly.

Figure 14.2 illustrates problems with an external stairway. The steps are open to the elements and get wet – leaf fall in the autumn adds to the slipperiness, and in winter they can get icy. Further problems related to dampness can be seen on the external wall of the building adjacent to the stairs and on the parapet wall.

Handrails and balustrading

These are important whether the stair be timber or concrete. Issues generally relate to the inadequacy of their fixings or their absence. A person climbing a stair, or holding on desperately to restrain him/herself from falling, can exert a powerful pull-out load on the fixing of a handrail or balustrade. Often, however, these fixings are poorly located or poorly executed. Checks should be carried out periodically and inadequate fixings attended to. This may mean from time to time providing additional or alternative means of support. Balustradings are particularly vulnerable at unsupported ends where they do not adjoin a wall; sometimes a member at right-angles to the panel can provide some worthwhile stiffening.

Figure 14.2 External stairs – unprotected from the elements, slippery and dangerous.

Means of escape; protected routes

Stairways are often significant contributors to safe escape in the event of a fire; they can also contribute to the spread of a fire. As far as possible, therefore, they ought to be provided within protected shafts which will contain fires and prevent their spread. This would entail their enclosure by walls with adequate fire resistance, and doors which are both self-closing and also of appropriate fire resistance. Management is also necessary to ensure that doors are closed, and not held open by fire extinguishers! Damage and unauthorised modifications to such provision should be observed and dealt with promptly.

Internal walls and finishes

These are multifarious and required to serve a number of functions. Masonry walls may be providing load-bearing capability to floors and further structure above (whether or not specifically designed so to do) and they may be self-finished fairfaced brickwork or blockwork, or they may be finished with plaster or some alternative. Non-load-bearing partitions may be of timber or metal stud and faced with a number of layers of plasterboard commensurate with the fire resistance required. The nature of the finish may be influenced by factors such as:

■ hygiene considerations, e.g. for a kitchen;
■ appearance;
■ cost; ease and speed of application;
■ fire performance and needs;
■ anticipated maintenance needs, including cleaning and decoration.

The BRE offers good advice on internal painting (BRE, 1998b).

Problems that may surface will generally relate to physical damage or to changing circum-stances. Plaster and, more so, plasterboard are vulnerable to damage; they are also fairly readily repaired, although redecoration may be needed in order to mask the joins. The British Standards Institution (BSI) defines standards for plastering (BSI, 2005). Occasionally cracks may appear, generally attributable to movement somewhere in the building; the cause must be identified and dealt with before repairing the plasterwork! It may be that the wall is inadequately supported, for instance built off a floor without a slab thickening to it or without a separate foundation; or it may be carrying an unexpected load. More rarely there may be problems with the plaster itself – perhaps if impurities have crept in, for instance in the water used in mixing, or by use of the wrong mix. That may require a wholesale replacement; a specialist investigation would be warranted.

Wall tiling can be problematic. There may be cracks relating to structural problems. There may be adhesion problems. Movement or moisture can force tiles off. As tiles are often laid in bathrooms and kitchens where excessive moisture may be present this can be a common issue. Adhesive technology has provided a wide range of products and it is a good idea to have the supplier give assurances (in writing of course) as to the appropriateness of the proposed adhesive for the fixing of the particular tiles to the particular substrate in the particular location and conditions. One has only to consider the consequences of a wrong specification in relation to swimming pool tiling to appreciate the importance of getting this right.

Summary

Having explored in this chapter the internal structure of the building, in the next we will look at what will, in many modern buildings especially offices, perhaps be the most expensive area of maintenance, the building's environmental services.

References

British Standards Institution (2005) *BS EN 13914-2: 2005 Design, Preparation and Application of External Rendering and Internal Plastering: Part 2: Internal Plastering.* BSI, Milton Keynes.

Building Research Establishment (1984) *Stairways: Safety of Users: Defect Action Sheet 54.* BRE, Watford.

Building Research Establishment (1997a) *Domestic Floors: Assessing them for Replacement or Repair – Concrete Floors, Screeds and Finishes: Good Building Guide 28: Part 2.* BRE, Watford.

Building Research Establishment (1997b) *Domestic Floors: Assessing them for Replacement or Repair – Timber Floors and Decks: Good Building Guide 28: Part 3.* BRE, Watford.

Building Research Establishment (1997c) *Domestic Floors: Repairing or Replacing Floors and Flooring – Wood Blocks and Suspended Timber: Good Building Guide 28: Part 5.* BRE, Watford.

Building Research Establishment (1998a) *Repairing and Replacing Ground Floors: Good Repair Guide 17.* BRE, Watford.

Building Research Establishment (1998b) *Internal Painting: Tips and Hints: Good Repair Guide 19.* BRE, Watford.

Construction Industry Research and Information Association (1998a) *Screeds, Flooring and Finishes: Selection, Construction and Maintenance: Part 9: Failure, Repair and Refurbishment.* CIRIA, London.

Construction Industry Research and Information Association (1998b) *Screeds, Flooring and Finishes: Selection, Construction and Maintenance: Part 10: Cleaning and Maintenance.* CIRIA, London.

Department of Communities and Local Government (2008) *Sulfate Damage to Concrete Floors on Sulfate-Bearing Hardcore: Identification and Remediation.* DCLG, London.

Society for the Preservation of Ancient Buildings (1999) *Care and Repair of Old Floors: Technical Pamphlet 15.* SPAB, London.

15 Defect recognition and rectification

Building services

The preceding five chapters have examined defects in relation to the external and internal structure and finishes of the building. This chapter looks at its related environmental systems, e.g. plumbing and waste services, heating and electrical systems. These will sometimes be very simple and rudimentary, little more than pipes and wires, but they can be very complicated and potentially hazardous if not given proper attention.

Professional bodies and organisations such as the Chartered Institute of Building Services Engineers (CIBSE, 2007), the Building Services Research and Information Association (Nanayakkara, 2003; BSRIA, 2008) and the Building Research Establishment (BRE) (Harrison & Trotman, 2000) offer valuable guidance in the field and reference should be made to this authoritative advice.

Something of the range of services provision and problems can be seen from Figures 15.1–15.4. Figure 15.1 is a photograph of the Lloyd's of London Building designed by Richard Rogers; it displays its service infrastructure on the outside of the building, leaving the office floors mush freer than they would otherwise be. Figure 15.2 provides a reminder that modern hospitals are highly serviced public buildings with much to maintain (including human life). Figure 15.3 illustrates how things look when they go wrong. In this example, cast-iron soil vent pipes have been made a prominent feature of the front elevation of this block of flats. Unfortunately the pipes have been leaking for some time, saturating and staining the brickwork behind and beneath. Tenants cannot be expected to necessarily notice let alone notify the landlord of matters related to communal features, so they tend to persist as problems. Figure 15.4 shows how buildings tend to be adapted over time; this Victorian house has been converted to several flats each with its own services arrangements – note the individual and idiosyncratic soil vent and boiler ventilation pipes.

Plumbing: water, waste and soil

Generally speaking, a building receives water via a pipe that is the last link in a series that has brought it from aquifers, boreholes or watercourses via reservoirs, water purification plants and a network of supply mains and service pipes. The building owner becomes responsible for the pipework from where it crosses the curtilage or at which it is metered. Normally that water is of potable (i.e. drinking water) quality even though much of it may be used for flushing toilets. The oldest water pipes were in lead – hence the word plumbing (from the Latin for lead); this is now recognised as toxic and injurious to health, including mental health. Lead plumbing should be replaced when possible. Some lead pipes are buried within floor structures and difficult to replace, however. Advice should be sought from the local water supply company.

Figure 15.1 Lloyds of London with its extensive external services provision.

Most buildings will have a cold water storage tank of a size sufficient to last for a period of temporary loss of direct supply. Most will also have a mechanism for heating water, for hand-washing at least, and maybe a hot water storage tank. I say 'most buildings' and 'normally' because there has been growing interest in trying to increase water efficiency, to reduce water needs and maybe use water collected on site. Robert and Brenda Vale for instance, in their 'autonomous house' at Southwell in Nottinghamshire gathered rainwater and filtered it through sand in large former wine vats to provide their own water supply (Vale & Vale, 2000). The building owner is responsible for all the water-related works on the site.

Occasionally there are other problems related to water supply pipes. They may be poorly fixed; there may be knocking or water-hammer heard where there is air caught up in the water flow, probably from an air vent to a hot water expansion pipe. There may be silty deposits of lime in hard water areas. There may be problems of electrolytic action where pipes of different metals are joined or at a hot water cylinder. Ball-valves can stick, resulting in overflows, often over entrances – they are there to be noticed and acted upon!

Most buildings also have pipes that take waste away from the building; by and large that waste is untreated. Water pipes and waste pipes often occupy related spaces. Water will be supplied from the mains and/or cold water storage tank to washbasins, baths, showers and kitchen sinks, and waste water will leave each of those appliances.

Figure 15.2 The twenty-first century West Wing of the John Radcliffe Hospital in Oxford.

As with water supply efficiency, there has also been growth of interest in reduction of use of water for waste transit. Low-flush or dual-flush w.c.s are an example, and fitting these should be considered when opportunity arises. The Vale's 'autonomous house' was supplied with a composting toilet from the outset. This dispensed with the normal waste system, replacing it with an appliance which turns human bodily waste to a dry material capable of being used to manure the rhubarb in the garden.

Principle problems with water and waste services relate to leaks. Clearly leaks must be attended to with urgency – the consequences can be very serious. Whenever a leak is detected the supply to the affected area should be turned off as soon as possible. On the whole leaks from water supplies, whether hot or cold, will be less problematic than those relating to waste systems; it is important to consider that what may appear to be related to the supply may be waste-related – it may not be obvious.

Once the source of the leak has been identified and remedied (whether by replacement of a pipe length or joint), the area is likely to need some drying out. This may be best achieved by ventilation, perhaps fan-assisted; it may take a while. Too rapid drying out may result in cracking of plaster or splitting of timber with consequential further work then needed in due course. If the leak related to waste water there may be grease to clean up if it emanated from a kitchen; if from a toilet then sterilisation may be required. It may be necessary to check for the presence of bacteria; this is a specialist activity.

Figure 15.3 Leaking soil vent pipes to a 1950s block of flats.

Figure 15.4 A plethora of pipes on a Victorian house converted to flats.

Very occasionally, but dramatically, there could be disease-related issues. If waste water were to leak into water supply systems or if affected people were to be involved, there could be an outbreak of diarrhoea or cholera. Vigilance is required to ensure high standards of design, installation and maintenance are upheld, especially in areas of the world where such problems are rife.

Gas

Gas is dangerous because it is explosive. In 1968, the whole corner of a 22-storey tower block of flats at Ronan Point in Newham, London blew out and collapsed due to a gas explosion in a kitchen. The gas had built up overnight and was ignited when a tenant put on the kettle to make an early morning cup of tea. Miraculously, only four people died, though that is tragic enough. The further consequences were huge: new homes had to be found for all the tenants of the block and people in similar blocks needed reassurance that they were safe in their homes. a programme of strengthening of system-built buildings all over the UK had to be developed and implemented. Millions of pounds needed to be spent; and faith in high-rise living was seriously dented. The cause was a faulty gas cooker connection.

Gas can accumulate in enclosed spaces. The build-up of gas in a duct under the floor of a three-storey block of six flats in Putney, London, resulted in eight deaths when it collapsed as a result of explosion (Hansard, 1985). Salford Council was fined £500 000 for failure in its duty of care to its tenants following a gas explosion in 2000 in which, mercifully, no-one was killed. The council employed a contractor to maintain the gas installation. This had not been done satisfactorily and gas had leaked into and filled wall cavities and thereby entered a flat, where it ignited and exploded when a tenant lit a cigarette (Inside Housing, 2008).

Ducts carrying gas pipes should be ventilated to the open air. Similarly, with appliances such as gas convector heaters, water heaters and boilers, there must be adequate fresh air supply to the rooms, including cupboards, in which these are located. People's lives are at stake, whether from explosion or from the build-up of toxic products of combustion. A programme of regular inspections should be instigated and visits recorded. Leak detectors are available and should be fitted; these should also be checked for effective operation. In Britain, all work carried out on gas installations must be carried out by qualified operatives of organisations recognised by CORGI, the Council of Registered Gas Installers.

Gas usage

We must be concerned to use less of all fuels, because they are limited in supply and because, linked to that, they are an expense. All gas use should be metered and consumption monitored. Peaks of use should be identified (times of day, week and year; and from one year to the next) and use in particular buildings and parts of buildings. The findings should be analysed, assessed and addressed. Energy is likely to be the highest cost of an organisation after staff. Particularly at times of recession, economic turndown, financial stress, reducing profits, people's jobs 'on the line', energy use must be curtailed and contained and be seen to be. The days of being able to reduce energy *costs* without reducing consumption, by shopping around for a better tariff, have passed.

There are also national and international imperatives. In rather the same ways that the western world felt threatened by the rising price of oil as demanded by the OPEC (Organisation of Oil Producing Countries) in the 1970s, governments of those same western nations and others are feeling vulnerable in relation to supplies of natural gas from Russia. The OPEC demands had encouraged the development of alternative oil and gas fields such as the North Sea; but this is now depleting. Energy security has risen up the political agenda. What this will mean in terms of energy supply is uncertain; perhaps it will mean that new alternatives and more self-sufficiency will become increasingly attractive.

Electrical and communication systems

Electricity is dangerous; it can kill by electrocution. Therefore regular inspections and testing should be carried out. Part P of the Building Regulations requires that electrical installations, including modifications and extensions to existing systems, be carried out by competent persons. As it is a specialist area, therefore, I do not discuss it in great depth here.

A rule-of-thumb has developed that the 'life' on an electrical installation should be considered to be 20–25 years. That was based on typical materials used in the 1950s, exemplified by rubber-coated twin-flex twisted fabric-covered wiring seen in lighting drops. The latter part of the twentieth century has seen huge growth in the development and use of electrical equipment in the home and at work. By and large it is the inadequacy of electrical networks within buildings (typically insufficient plug sockets) that has prompted upgrades in electrical systems rather than technical failures. Appliances that were once considered luxuries are now necessities: toasters, refrigerators, washing machines, dishwashers, televisions, computers, portable appliances including electric kettles, carving knives, electric drills and other power tools, etc.

A facilities or maintenance manager may be responsible for the maintenance of some of these items of equipment, and for more specialist facilities such as in a commercial kitchen or workshop or laboratory. They will certainly be responsible for supporting electrical infrastructure. Programmes of regular inspection and testing will need to be developed and implemented, with good and reliable systems of recording put in place and maintained.

Technology keeps moving on. Moore's Law suggests that computing power doubles every 18 months. Continuous, or at least continual, upgrading must be expected and as far as possible planned for. Some developments, like the growth of IT in business, have rendered some buildings difficult to use, outdated, old-fashioned, redundant and hard-to-let because they could not accommodate or be readily adapted to the substantial amount of wiring required for the new systems. A lot of businesses moved from 1950s offices with limited floor-to-ceiling heights and cellular offices to new buildings in the booming 1980s. These new buildings were typified by open-plan offices with access floors, to accommodate the miles of wiring to serve a personal computer on each desk, and air-conditioned with ductwork hidden above suspended ceilings (more on this below). Developments of wireless technologies in the early twenty-first century have rendered a lot of the access-floor approach redundant, and are making it possible to breathe new life back into those 1950s buildings that are still standing!

Energy security was referred to in the section on Gas, above. The increasing price of electricity has prompted a renewed interest in nuclear power, however unattractive and unpopular that may be. Greener alternatives may become more readily available and competitive; it is possible to sign up for a 'green tariff' that favours 'greener' modes of power generation. Self-sufficency may become possible where electricity can be generated on site (by solar panels, water or wind driven turbines, etc.) and this matched to the level of demand of the building. This may be feasible and indeed desirable for relatively small enterprises in rural environments. 'Microgeneration' in such ways will be harder to achieve in urban areas.

The UK government issued Planning Policy Statement 22 in 2004, which set out guidance in relation to policy regarding renewable energy (Office of the Deputy Prime Minister, 2004). The London Borough of Merton responded enthusiastically to this guidance developing what has become known as the 'Merton rule'. Developers were required to use on-site renewable energy resources on all major new developments wherever viable. Other local authorities have developed similar policies. The cabinet member for housing and property at Kirklees Council (an urban area on the edge of the Pennines) has been quoted (The Merton Rule, 2008) as saying: 'We are effectively setting our own version of the building regulations . . . '; and therein lies the rub. Building Regulations are intended to be consistent across the country, so this kind of local variation is unwelcome to developers who operate nationally and they have questioned and challenged the validity of the 'Merton rule'. There is certainly no issue in relation to building owners exceeding the requirements of the Building Regulations and many organisations are seeking to develop their 'green credentials' as part of their corporate social responsibility. Such plans may be expected to feature in companies' annual reports, so it may be beneficial for facilities managers to be, and be seen to be, proactive in this area.

Heating, ventilation and air-conditioning

These are problematic areas; potentially high consumers of energy, rich sources of discomfort, dissatisfaction and complaint, and dangerous. Potential problems include sick building syndrome (SBS) and legionnaires' disease.

What is SBS? Symptoms are fairly well known but the cause(s) have been harder to pin down. Rostron (1997) provides a good overview. Headaches, dryness, irritation (and irritability), runny noses and eyes and lethargy are common components. Air-conditioning, poor ventilation, dust mites, background noise and/or vibration are cited as possible contributors. A cynic or sceptic may suggest it is a kind of whingers' disease, implying that is an expression of dissatisfaction with aspects of an individual's working conditions which may not be related to the building at all; it is a label that is OK to use. All complaints or concerns should be taken seriously and addressed as far as possible. Bearing in mind that asbestosis and mesothelioma were unknown and denied for some time but have become common matters of concern in relation to litigation and liability, it is probably wise to respond. It is good practice, of course, to be concerned for people's welfare and, in any case, building owners and occupiers owe a legal duty of care to all who use their buildings.

Legionnaires' disease is so called because it was first identified when an outbreak of pneumonia occurred at a convention of the American Legion in Philadelphia in 1976. This was subsequently determined to have been caused by the bacterium *Legionella pneumophila*. The bacterium thrives in warm water and can be dangerous when found in shower heads or whirlpool baths where spray can give rise to airborne particles that may be inhaled. Water-cooled air-conditioning systems are particularly problematic, as from the rooftop location of a water cooling tower, spray can be spread over a wide area. A council architect, Gillian Beckingham of Barrow-in-Furness Borough Council, was fined £15 000 for breaches of health and safety law, and her employer £125 000, when seven people died from an outbreak of *Legionella* caused by lack of maintenance of the air-conditioning system of the municipal arts centre. Hotels and leisure facilities are particularly vulnerable. Regular inspections, testing and cleaning are required. The Department of Health (DOH) has issued standards (DOH, 2006).

Air-conditioning ducts are another source of potential problems. Dust and particles of any sort of airborne pollutant may be present and ready to be circulated to any part of the building. Regular cleaning is essential. Ducts can also be irritating in relation to noise and vibration.

The thin metal walls of the ducts can act like a drum when vibration is introduced, perhaps from a fan; and such noise can be very irritating. Ducts are also a mechanism by which sound can be transmitted from one room to another (not good if it is a conversation that can be overheard) and, more dangerously, through which fire and smoke could be spread. Ducts and pipes can be points of weakness where they pass through structure that should be preventing the passage of fire. Fire-stopping and dampers, perhaps with intumescent materials, should be used to shut down these apertures in the event of a fire. Areas unseen above suspended ceiling – 'out of sight; out of mind' – are particularly vulnerable and must be checked periodically.

Controls on heating and air-conditioning systems also need periodic checking. There can be a tendency for settings on thermostats and other control devices to 'creep' over time. Thus services can be on for longer than needed or intended, or at higher temperatures before switching off, cutting in at lower temperatures, or cycling between highs and lows. Reducing maximum room temperatures by a degree or two can make a substantial reduction in winter heating bills, and accepting higher temperatures can reduce summer cooling bills. Metering, monitoring and making changes in response to the measurements are key components of a maintenance regime.

The possibility of manual over-ride of corporate systems and standards may be thought to offer the possibility of mayhem and huge bills, but evidence suggests that occupiers are rational and not reckless, and that giving people control over their own working environments is in fact cost-effective. Leaman and Bordass (2000) found that 'people were more tolerant of conditions the more control opportunities – switches, blinds and opening windows, for instance – were available to them', and that 'high perceptions of personal control bring benefits such as better productivity and health'. The Department of Communities and Local Government (DCLG) has issued guidance on inspection of air-conditioning systems in relation to improving the energy efficiency of buildings (DCLG, 2008).

Fire, security and other specialist services

Fire protection measures may be broadly described as *passive* or *active*. Passive measures are those such as compartmentation by fire-resisting structure and protected escape routes. Active components could include automatic fire sprinklers and smoke vents.

All components that play a part in resisting and containing the spread of fire and allowing a safe means of escape should be inspected regularly: walls, doors and closers, corridors and stairways, fire blankets and extinguishers, smoke and heat detectors, alarms, emergency lighting, escape routes and signs, sprinklers, smoke vents, suppression systems and so on. Failure to do so is negligent and may lead to loss of life.

Times at which to keep such matters particularly in mind include:

■ departure of key staff; lack of awareness and training of new staff;
■ reorganisations of uses of space, intensifications of use, potential unintended restrictions of escape routes;
■ redecorations, resulting in use of materials with potential to spread flame, e.g. wallcoverings, gloss paints;
■ execution of maintenance or construction-related work resulting in equipment or materials in escape routes;
■ fire doors propped open, for example for furniture removals;
■ storage of volatile materials, e.g. cleaning products.

Security

This is an area of provision that has grown over the years in response particularly to the terrorist threat. In UK, attacks by the Irish Republican Army from the 1970s and from Al Qaeda and followers of Osama Bin Laden after the outrage against the Twin Towers of the World Trade Center in New York on 11th September 2001, and subsequent terrorist outrages in London and Madrid amongst others, have encouraged a defensive attitude. In extreme cases, such as around the Houses of Parliament, this has given rise to substantial (and ugly) physical barriers of concrete blocks; the home of the Prime Minister in Downing Street is fenced in by substantial railings and guarded gates. Many buildings are protected by closed circuit television (CCTV) systems where cameras can be remotely operated and monitored. The US Federal Emergency Management Agency (FEMA) has prepared advice on the protection of commercial buildings against terrorist attack (FEMA, 2008).

All security systems need to be routinely checked, not only to ascertain that the mechanical parts are operational, but also that the human interface is working. Jobs in the security industry are notoriously poorly paid, and the work can be numbingly boring. It can be very hard to stay alert, or even awake, through long night-time hours in which nothing generally happens. Security staff must be vetted for their trustworthiness and reliability.

Increased interest is being shown in the application of biometric recognition systems, by which people's eyes, faces, fingerprints and/or palmprints are recorded and compared within a database. There is concern about the security of such databases but discussion of such issues is beyond the scope of this book. Clearly the maintenance of such equipment is vital to its accuracy and availability. Access control that is dependent upon such systems will need to be risk-assessed and have well developed and reliable contingency plans ready to swing into operation. Such standby systems must be fail-safe such that malign miscreants and malcontents continue to be denied access while exit is not impeded to people trying to escape a building in which a fire has been detected.

The more sophisticated or complex the system, whether technical, organisational or administrative, the more likely it is to be problematic in its operation and perhaps in its rectification when things do go wrong. Appropriate systems need to be in place for all foreseeable eventualities, including the unforeseeable (Levitt, 1997). It must be foreseen that although their detail will be unknown, unforeseeable events will occur; they too must be planned for. Periodic brainstorming may be a useful maintenance team exercise, with vigilance in relation to assessment of and action upon the perceived level of risk.

Summary

This chapter completes our series of six chapters dedicated to the examination of defects in the structures and services of our buildings. However, our buildings' and occupiers' needs do not stand still. Standards and expectations keep changing – this is the subject of the next chapter.

References

Building Services Research and Information Association (2008) *Maintenance for Building Services: Building Application Guide BG3/2008*. BSRIA, Bracknell.

Chartered Institution of Building Services Engineers (2007) *Inspection of Air-Conditioning Systems*. CIBSE, London.

Department of Communities and Local Government (2008) *Improving the Energy Efficiency of our Buildings: a Guide to Air-Conditioning Inspections for Buildings.* DCLG, London.

Department of Health (2006) *Water Systems: HTM 04-01: The Control of Legionella, Hygiene, Safe Hot Water, Cold Water and Drinking Systems – part B: Operational Management.* DoH, Leeds.

Federal Emergency Management Agency (2008) *Incremental Protection for Existing Commercial Buildings from Terrorist Attack: Risk Management Series 459.* FEMA, Jessup, Maryland.

Hansard (1985) http://hansard.millbanksystems.com/written_answers/1985/jul/25/putney-gas-explosion. Accessed 25 February 2009.

Harrison, H. & Trotman, P.M. (2000) *Building Services: Performance, Diagnosis, Maintenance, Repair and the Avoidance Of Defects.* BRE, Watford.

Inside Housing (2008) www.insidehousing.co.uk/story.aspx?storycode=6500586. Accessed 25 February 2009.

Leaman, A. & Bordass, W. (2000) Productivity in buildings: the 'killer' variables. In: Clements-Croome, D. (ed.) *Creating the Productive Workplace.* E. & F.N. Spon, London, pp. 167–191.

Levitt, A. (1997) *Disaster Planning and Recovery – A Guide for Facilities Professionals.* Wiley, New York.

Nanayakkara, R. (2003) *Guidance for the Procurement of Building Services: Operation and Maintenance.* BSRIA, Bracknell.

Office of the Deputy Prime Minister (2004) *Planning Policy Statement 22: Planning Guidance on Renewable Energy.* ODPM, London.

Rostron, J. (ed.) (1997) *Sick Building Syndrome – Concepts, Issues and Practice.* E. & F.N. Spon, London.

The Merton Rule (2008) http://www.merton.gov.uk/living/planning/planningpolicy/mertonrule.htm. Accessed 27 February 2009.

Vale, R. & Vale, B. (2000) *The New Autonomous House.* Thames and Hudson, London.

16 Upgrading and improvement

The six preceding chapters have dissected and discussed defects across a range of building elements, looking at what has gone wrong and why, and how they may be corrected and be avoided in future. These will represent a lot of the activity of a maintenance manager. However, to merely maintain the *status quo*, or restore the *status quo ante*, is to ignore a great deal of his or her potential input and to fail to realise the potential of the building.

This chapter therefore takes a more future-focused view.

Changing standards

The word 'upgrading' suggests improvement; a change or changes to the building that brings about a significantly higher level of performance than previously; better than that which the building achieved at the outset. Another word that could be used would be modernisation, but now that we have postmodernism it doesn't have the same connotation.

We have noted how buildings, their component parts and the service they offer deteriorate over time. In most circumstances, a building will reach a point where it fails to meet the need in some significant regard. How might that point be identified and what should then be done? An example is given later in this chapter that relates to deciding between effecting changes *in and to* the building as opposed to a change *of* building, i.e. moving.

It is helpful to be able to anticipate when such decisions might need to be made. A building manager who knows his or her buildings and what is happening in and to them is more likely to have a sense of when that time is coming. Data may also help, if it is appropriate. For instance, if a particular kind of repair or replacement activity is becoming very frequent, or if there is significant activity in some part of the building, these may be prompts for intellectual activity to consider options.

There are also changes outside the building itself that may influence or even determine a need to change. Reference was made in Chapter 1 to 'PEST' amongst a range of tools for considering the external environment in which our buildings function. Standards result from and represent a combination of factors, including political, economic, social and technical. For instance the development and imposition of higher standards of energy efficiency stem from:

- increasing price of fuel;
- a wish to 'save the planet' and for everyone to assist in and benefit from that;
- technical ability to provide more energy-efficient forms of construction;
- political will to impose the higher standard in order to reduce energy use and increase energy security.

What determines the appropriate standard to set? Usually this is a compromise between what is technically and economically feasible. The UK government has developed a rolling programme of review of statutory controls, such as Building Regulations, so that designers have time to familiarise themselves with and adapt their designs and practices to the new or changed standards. Of course statutory levels of performance must be considered the minimum acceptable. Unfortunately, there is a tendency for these intended minima to become maxima; I don't know why that is. For instance it may make economic as well as environmental sense to upgrade thermal performance above and beyond that required by Building Regulations, but often the stated standard becomes the target.

Equally, but more unusually, it may be that standards required will be reduced over time. For instance it might be thought appropriate that the standards required of a building's performance should be expected to reduce over time – a kind of recognition of ageing rather like we might expect a person to 'slow up' as they become elderly. From time to time, reviews of standards suggest that some areas of performance that were deemed necessary to be controlled at some time in the past are no longer thought to be in need of that control. For instance, Building Regulations used to specify minimum storey heights and to make demands of zones of open space outside openable windows in order to ensure adequate natural ventilation. These controls were completely swept away in the 1985 revision of the Building Regulations, which also re-presented how all the regulations were framed and phrased.

Incremental change

Buildings can be upgraded every time some repair or replacement is implemented. Each improvement may be so small as to be barely noticeable as an improvement, and the rate at which they are introduced may be such that, even though in total over time they amount to significant improvement, memory is so poor that the improvement is unnoticed. This is also a measure of how increased standards over time are almost expected and taken for granted.

Such incremental change can be introduced almost imperceptibly into the building and with little or no effort or expenditure above that of a like-for-like replacement. It will generally be a matter of identifying the current model that is broadly equivalent to the original. Most manufacturers have policies of continuous improvement of their products, increasing performance of the 'new improved model' over the original model and/or adding functionality. Sometimes this policy can be a nuisance in that replacements have marginal differences from the original that give rise to the introduction of visible variety where previously there was consistency in appearance.

Unacceptable variation of this kind can be avoided, reduced or at least deferred by the carrying of spares bought as part of the original construction package. In fact often the construction procurement processes are such that spares are produced by the purchase of items in multiples of say ten or a dozen and by over-ordering to allow for damage and other wastes. There is then the problem of storing these spares somewhere and of remembering that they are there when a replacement is required! It may make sense to arrange to keep spares of some of the more significant items where matching may be important. However it must be remembered that elements on the external surface of the building will deteriorate and discolour over time so blending in an 'as new' component amongst weathered items will be difficult.

It may be better to accept that new items will look different and to discipline oneself to appreciate the honesty of the new introduction. The Society for the Protection of Ancient Buildings (SPAB) promotes this approach. Thus, for instance, a piece of decayed brickwork may be rebuilt using clay tiles to accentuate the change; spalled stonework may be replaced with the face of the new stone aligned with the original surface rather than that of the decayed areas that remain, again highlighting the change. This may be seen by some perhaps as incremental visual degradational change.

Exponential change

Some people say that the rate of change is increasing. This is hard to measure, but a postmodern way of looking at this would suggest that if that is how it feels then that is how it is. Is a change of standard required of a particular item from A to B and then at a later date to C more change than if the standard of one kind of item is changed from L to M and that of another from X to Z? On the whole people do not welcome, let alone embrace, change. So comments related to accelerating change are not generally extolling its virtue. Of course it is axiomatic that the more one experiences change the older one is getting so it is not a surprise that on the whole more older people are more conscious of and resistant to change. And because we cannot return ourselves to the situation before change nor divorce ourselves from the effects of change, we can hardly be objective when suggesting that 'things are not as good as they were'.

Exponential change could be effected by having change happen increasingly frequently or by having it happen at regular intervals but by increasingly large amounts. Substantial change can be achieved fairly quickly by making relatively small changes at intervals over a period. The effect is the same as with compound interest. A change of 7% per annum over 10 years will produce a doubling of the previous performance. That sounds like something worth striving for and attainable. However, while it may be relatively easy or straightforward to make the first 7% change from perhaps a very poor previous situation, it may become increasingly difficult to achieve similar gains in future years. Also, is it reasonable to make interventions at that kind of interval? While some people may like that kind of demonstration of continuing interest and attention, it may feel like a continual tinkering – a job that is never-ending. Some may prefer a kind of 'big bang' once-and-for-a-long-time approach to change – larger and less often.

Step change

This kind of change has a lot of attraction, especially to those who like to see impact. It is very appropriate when buildings have become run-down and are very much 'below par'. A step change has the advantage of being transformative; the improvement, and the scale of it, are obvious. It has strong PR (public relations) potential; 'before' and 'after' photographs can be taken and used in promotional literature. It looks like the organisation is 'moving ahead'.

However, such change may be expensive. The scale of change may be such as to require substantial parts of buildings, maybe entire floors, or whole buildings, to be vacated. The length of time needed to effect this can be substantial; there are a number of things to be considered, such as:

■ What change is required? How will this be decided? Who will be involved?
■ How long will the building work take?
■ How long will it take to pack up and move to vacate the area?
■ Where will people be accommodated during the work?
■ Will they move back; if so when; and how long will that take?
■ How long will it all take in total?
■ How will the disruption and loss of production be handled?
■ Is it worth it; what are the alternatives?

A particular advantage of a step change is that it encourages a radical review of activities:

■ Why are things done the way they are? Why are they done at all?
■ Where should they be done; how much space is needed; and of what kind?
■ Who decides? Who should decide?

A step change also suggests that it will not happen very often, although it may. It will probably be hoped that due to the scale of change, it will not happen often. People will look forward to moving to the changed state (albeit with some trepidation) after a period of disruption; they will not look forward to being 'messed about with' again shortly after or repeatedly. In that people will be expected to go through a process of accommodating themselves (hopefully with support) to the new environment whatever the scale of change, it is appropriate to consider making the largest change that is supportable.

Assessment of the appropriate changed state may include:

■ What is the 'best in class' currently achieved? Can that be achieved here?
■ Could that be improved upon; by how much; at what cost? Would it be worth it?
■ How to get from 'here' to 'there'?
■ Are the people 'up to it'; 'up for it'?
■ What support will be required; what is available?
■ What funds are available?
■ How long before the next change may be expected?
■ What will happen if no change is made now?

The bigger the change the more important it is to get it right (or at least close to right), and to apply appropriate management time and effort to ensure the change goes well – both the process and the end-result (as much as there is a recognisable *end*). So planning and monitoring are essential. Feedback is also important to inform future changes in terms of process and product.

Consideration will also need to be given to what is to happen after the step change, if the higher standards are to be maintained. Perhaps a longish gradual decline in performance may be considered acceptable. Alternatively it may be hoped that staff now accustomed to the high standards may have had their expectations raised and a programme of future incremental improvements may now be appropriate. Staff may become dissatisfied and restless, perhaps even leave the organisation, if it fails to 'keep up'. And staff that work for organisations perceived by their competitors as being leaders, in this case in terms of workplace, will be seen as attractive to rivals.

Figure 16.1 is intended to illustrate the periodic introduction of higher standards and their relationship with performance of the whole building or of any of its various elements. It shows in a simplified and smoothed diagrammatic form how buildings need to be periodically upgraded in line with changing (generally increasing) expectations.

Demands and expectations keep changing. In the UK a lot of work was done in the 1990s by the then Building Research Establishment Energy Conservation Unit (BRECSU) to promote increased energy efficiency in Britain's buildings. Good Practice Guides and Case Studies were published related to most major building types – new and existing housing, schools, factories, offices, theatres, hospitals, universities and public houses.

More recently there has been a good deal of interest in the concept of urban regeneration, with the aim of bringing new life back into old and rundown towns and cities that were previously centres of industry: manufacturing, mining, shipbuilding, steel-making, weaving and so on. In some areas this has given rise to market renewal programmes, with plans to demolish some

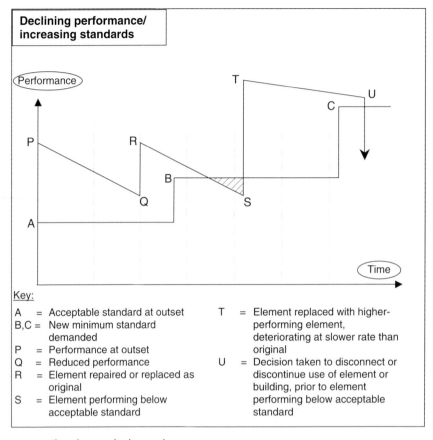

Figure 16.1 Changing standards over time.

terraces of worn-out and substandard housing while upgrading others. This coincides with a belated recognition that Britain's energy use will only be brought under control by major upgrading of its older housing – a step change. The Building Research Establishment (BRE) and the Sustainable Development Commission (SDC) are at the centre of this thrust (SDC, 2006; Yates, 2006a, 2006b; Plimmer *et al.*, 2008), setting out methods for assessment and benchmarks. A good example of work at the local scale is available as a compendium of 15 case studies of sustainable housing refurbishment (Hastoe Housing Association, 2004).

The Chartered Institution of Building Services Engineers and Building Services Research and Information Association (2007) have set out to promote energy-efficient refurbishment more generally.

Local or international standards

Another factor in deciding the appropriate standard to apply is the local environment or context. An organisation which functions only in and around Liverpool, for instance, may feel it only needs to consider standards appropriate to 'holding their own' against competitors elsewhere in Merseyside. A larger and more diverse organisation, though similarly located, may give consideration to working conditions enjoyed further afield in the northwest of England, for instance in Chester, Manchester and Preston. If it wishes to set up some operations in Wales or Scotland there may be national differences; for instance, building standards differ in Scotland. International operators will need to consider the differing standards, and rival attractions, of London, Frankfurt, Paris, New York and Tokyo, to name just a few.

Multinational organisations may feel it appropriate to apply one set of standards across all its operations, perhaps irrespective of national boundaries and cultural differences. Sometimes this can come over as a kind of homogenisation, or worse of 'cultural imperialism' where the standard of one, the 'home' country, is imposed in and on others. This may stem from powerful ideas of branding such that the company's product is the same the world over, or perhaps of consistent processes such that all regions report to HQ in a way that enables summations, summaries and comparisons to be made readily. It also facilitates mobility and deployment of staff.

The growth and success of Japanese industry in the late twentieth century prompted much interest in cross-cultural understanding and practice (e.g. Geertz, 1973; Hofstede, 1980; Trompenaars & Hampden-Turner, 1997). Not only is it important to understand differences in culture of potential purchasers of the organisation's product or service, but also of the organisation's own workforce. For instance, whereas it may be generally acceptable for an office worker somewhere in the UK or USA or Australia to remove a jacket if they feel too warm at work, it may be necessary in another culture to seek permission. Somewhere else it may be thought inappropriate to seek such approval due to concern about being thought of as troublesome or questioning of authority. These behavioural norms may affect the way heating, cooling and ventilation systems are designed, installed and controlled.

Standards in the formal, official, authoritative sense have tended to be set by national bodies, such as the British Standards Institution in UK. Increasingly, with a wish to reduce barriers to trade there has been a trend toward common international standards. The European Union and its predecessors have promoted the development of European Norms and there is a growing body of International Standards. Thus many British Standards, for instance, have been developed or accepted for wider application with joint numbering. Thus, in relation to quality management systems, what started out as BS 5750:1979 became dual numbered as BS 5750 EN 29000 ISO 9000 in 1987.

Of course it is difficult, and I suggest inappropriate, to develop and apply common standards of *technical* performance around the world; the climatic conditions encountered are so different. For a building or component to be able to perform adequately, let alone to its best, wherever it is located would be an expensive enterprise. Equally it is little value to develop a set of 'lowest common denominator' standards – this would result in accepting, if not promoting, a lot of not very good practice. What is required is to apply *at least* the national standard and to consider how much that should reasonably be improved upon.

What do we want?

Do we want our buildings to be the best they can be; or do we want them to be just good enough to do the job; or some intermediate position; or something else? What are the motivations for upgrading? What will be the determining factors? Where to start these considerations?

The starting consideration must be the building as it currently is. How well does it meet current standards? How much 'headroom' is there between current performance and today's minimum acceptable standard? At current rates of deterioration, when will some part of the building fail to meet some current standard? What changes in standard are already 'in the pipeline'; and what further changes can be anticipated? Thus when will the deteriorating building fail to meet the future standards? That must be the minimum upgrading required, unless of course it is decided that the building is not worth upgrading; that possibility is considered further later. If the building is to be upgraded it is necessary to decide by how much as well as when. The building needs to comply for as long as it is in occupation or operation. It will be particularly disruptive if non-compliance occurs while the building is in use and operations have to be suspended and relocated pending improvement of the building, so timeliness is important. Bumbling along just above, or below, the limit of acceptability is tiresome, tiring, time-consuming and unrewarding professionally.

It may be thought that to function at a level higher than the minimum required is to waste money; but the situation explored above is expensive too in terms of continual intervention and expenditure to more or less stand still in a near-unsatisfactory state.

Something like an annual and/or a quinquennial (5-yearly) review is about the right interval – shorter intervals and you never 'break out' of the immediate; longer and it becomes difficult to project ahead with any accuracy and/or too many unforeseen matters will have intervened. This does not mean yearly or 5-yearly upgrading, only review of the situation, in order to determine what, if any, upgrading may be warranted.

Similar organisations

Another factor in determining appropriate standards may be what similar organisations or competitors may be doing. In some spheres, particularly in the public sector, it is common to 'compare notes' with similar organisations. Sometimes these will be adjoining geographically, for instance district councils within the same county; other times it may be more appropriate to consider comparators such as fellow boroughs in other metropolitan areas. This can be rather like 'town-twinning', trying to identify similarities of important aspects such as size and distribution of buildings by age and type. While of course it is impossible to find exact equivalents, the aim is to minimise the range of underlying differences so that the effects of variations in practice can be discerned and considered. It takes a certain professionalism and inner confidence to submit one's performance to such scrutiny; this is hard if one expects criticism and condemnation as a result. A climate of supportive consideration is a prerequisite.

In the private sector, competition rather than collaboration is more the order of the day. Challenge to existing practice tends to come with staff recruited from competitors who are able to make useful comparisons and proposals for change. Some organisations are operating in markets where price is crucial; for them it may be difficult to apply standards to their buildings other than those which will drive down costs. For others, where quality of service is key, and especially for premises visited by customers or potential customers, higher standards may be appropriate, perhaps even determinative of whether a potential customer is 'converted'. Someone will need to decide whether to apply higher standards to only those areas visited by customers – for instance the entrance foyer and meeting rooms (and access thereto) – and thereby run the risk that the workers feel devalued, or to not differentiate.

A commercial decision can be taken on standards:

- What can be afforded?
- What must be afforded; what is necessary to do the job?
- How will the result compare with 'competitors'?
- Can expenditure be deferred; should it be brought forward?
- Can differential standards be applied?

■ Will staff leave, or be attracted?
■ Will customers leave, or be attracted?
■ How do the proposed standards fit with the organisation's values?

How to decide?

As suggested above, the answer you reach on what is the appropriate standard to apply depends greatly on the criteria you apply; for instance:

■ lowest first cost;
■ best we can afford (however defined);
■ best in the long run (how long?);
■ whatever staff want;
■ what I like;
■ that which will bring us most business;
■ what will look good in the photographs/press/media coverage.

If for example, you are very financially orientated, you will probably be much influenced by cost considerations; if you are a 'people person' you may be persuaded by what staff say they need, or you may put the customer first. The 'balanced scorecard' approach (Kaplan & Norton, 1992) may have value here. Those affected, or more accurately perhaps, those involved with deciding, should allocate proportions of points to the various criteria and then mark against each criterion, adding the total score.

If time and/or money permitted, or it were thought imperative in promoting staff involvement, it may be possible to develop alternative 'sketch' proposals of change and to submit these to evaluation. Participants will prefer one proposal to another depending upon their criteria and how well the proposals 'measure up' against those.

For an organisation with a number of buildings it may be a good approach to assess what may be considered the best of the buildings, and to endeavour to match or better that standard. Of course again what makes for the best will depend upon the criteria applied. It is good practice to keep the performance of buildings under regular review, so there should be already agreed criteria and ratings; if not, this is a good time to start. It may be that a building will perform better than others on one or more criteria, and less well on others. It will then be necessary to try to draw on the best from several buildings and/or the best overall. It will also be possible to ascertain how colleagues within the organisation perceive matters.

How far to go?

Standards can almost certainly not be improved without expenditure. How much is worth spending to achieve any particular standard? There is a number of methods available for helping to determine an answer. There may be several answers, again depending on the values put into whatever equation is used. Assessments can be based on such as discounted cash flow (DCF), net present value (NPV), return on investment (ROI) and return on capital employed (RCE), which try to compare the increased value with the amount of investment needed to produce it. There is much to be said for a simple calculation of the payback period after which the improved building or service is making money for the client, having repaid the investment. It generally takes no account of interest charges or of enhanced or depreciated capital values as they tend to complicate the calculation and introduce more scope to 'fiddle'. On this simple payback calculation, various potential investments can be evaluated. Consider the following possibilities:

(a) Saving in running costs of £50 000 p.a. from investment of £0.6 million: payback 12 years.
(b) Saving of £100 000 p.a. from investment of £1 million: payback 10 years.
(c) Saving of £20 000 p.a from investment of £20 000: payback 12 months.
(d) Saving of £10 000 p.a. from investment of £50 000: payback 5 years.

Generally items with a payback period of less than 2 years would be fairly uncontroversial, and those of up to 5 years very attractive; beyond that it may be harder to justify finding the money now but if it is available it would be good to consider further. In the above examples, (c) and (d) could be worthwhile, and (a), although it is a longer payback than for (b), may be worth considering in that the amount of investment is less than for (b). Some simple sensitivity analysis would be worth applying, bearing in mind that the investment required may be greater than expected and/or the savings may be less; they are hard to assess with any great accuracy. It may be possible to negotiate with a contractor to 'guarantee' a certain amount of improvement for a given investment, although clearly the contractor will allow for the risk in determining the price to offer.

Such calculations are certain to be hedged around by uncertainties. It may be that an organisation will be attracted to the idea of moving to a new building that is considered capable of meeting the higher standard required without the risks attached to improving an existing building.

Move or improve?

'Better the devil you know': that is an expression that suggests there is much to be said for staying put, where you know what works well and what doesn't, rather than changing to a relatively unknown situation, where you have yet to find out what doesn't work. It may sound like a rather negative way of looking at things, but perhaps worth having in mind as a note of realism when considering alternatives.

For those pursuing a policy of continuous improvement and relatively small, if frequent, incremental changes, a building can almost certainly be adapted to accommodate the desired changes. However, it may be that over time a number of small inefficiencies or inconveniences consequent upon or required to accommodate to the changes are tolerated that might not otherwise be thought good or acceptable.

Where a step change is required, this will be difficult to accommodate without major disruption. If the scale of the operation is such as to warrant moving out for the period of the works it may be worth considering a permanent move, whether to a newly built building or to another building converted specifically to the organisation's new needs.

For instance, it may not be possible to keep premises in operation while dusty and/or dirty building operations are in progress. Cleanliness, hygiene, noise and security may be issues. In buildings with substantial public access, business may be seriously disrupted and lost. It used to be a 'no-no', for example, for a public house to close for refurbishment for fear the lost trade may not return. This is not so much the case today, as often part of the rationale for the work is to change the image, lose the existing, uneconomic, 'regulars' and replace them with a new free-spending clientele.

The foregoing may not be so much an example of a choice between staying put and moving, but it does illustrate the need to relate continuity of business operations to scale of building works. In essence, the larger, longer and more disruptive the work, the greater the need to contemplate a long-term, perhaps permanent move.

Possible advantages of a permanent move may include:

- new, leading edge, equipment and technologies;
- clear indications of 'moving on' and 'moving up';
- ready sign of confidence: 'investing in the future';
- good feeling amongst staff;
- a 'fresh start';
- opportunity to configure processes, workgroups etc. to maximise quality of output;
- location to suit needs of today and tomorrow, rather than yesterday;
- only need to move once (no moving back);
- staff can get up and running quickly rather than working out of boxes for duration of the temporary move.

Possible advantages of staying put, and upgrading there include:

- continuity;
- staff commuting patterns are based on current location;
- customers and suppliers are related to current location;
- smaller scale of financial investment;
- ability to phase changes;
- ability to adapt plans in the light of experience or of changed circumstances.

Now these simple lists above may suggest that there are more (potential) advantages in moving than in staying put, but that is to suggest that all will apply in every circumstance, which they will not, and that all are of equal weight, which also they will not be. The one word 'continuity' for example may represent a whole lot of issues related to well established ways of doing things, 'comfort zones' for people, distrust perhaps of untried and/or potentially unworkable new ways or systems. So again some kind of weighting and rating exercise may be appropriate. For instance, for some people and organisations a long-established connection with a particular place may be important. It may be that 'this is the place where we started it all'. For instance the chocolate makers Terry's of York opened their factory there in 1886 and moved to a new factory in York in 1926. However, following purchase of the company by Kraft Foods in 1993, production was moved to Eastern Europe and the company's presence in York came to an end in 2005.

The University of Oxford needs to be in Oxford – alternative locations are unthinkable. Oxford Brookes University – its junior counterpart, formerly Oxford Polytechnic – considered relocating in the 1990s but has remained in Oxford, albeit in four suburban locations.

A more generic connection may be important – for instance the association of steel with Sheffield. The Company of Cutlers was set up in that city in 1624, and people will still associate Sheffield with high-class cutlery today. There is value to a cutlery firm to be located in Sheffield.

Mini case study

An example of a phased relocation is that of the John Radcliffe Hospital in Oxford.

The Radcliffe Infirmary opened in the centre of Oxford in 1770. The old buildings were incapable of being brought up to modern, twentieth century, standards and meeting today's requirements, let alone those of tomorrow.

A substantial new building was opened on a larger, less constrained site in suburban Headington in 1972, and a sizeable new wing and Children's Hospital were added in 2006. Relocation

of all functions from the Infirmary to Headington allowed the closure and sale of the old site to the University of Oxford.

This brought in substantial funds while at the same time giving Oxford a modern hospital with all functions together on one site.

Checklist

Each situation demands consideration in its own context, so it is hard to generalise even about the questions to be asked. I offer the following as a very general starting point rather than as a fixed template.

1. What are the jobs needing to be done by the organisation?
2. How much space do they need?
3. What adjacencies are needed – i.e. what needs to be located next to each other?
4. Can these be accommodated in the existing building?
5. How much space is left over/surplus/spare?
6. Can adequate arrangements be made for circulation routes, emergency exit, toilets, lighting (daylight and/or artificial), sunlight/views, access/deliveries/security/parking, other, . . . ?
7. Have you thought about storage?
8. Space for future expansion?
9. What does the future hold? What range of possibilities can be accommodated?
10. What alternatives have been/could be investigated?
11. What is on offer elsewhere; how does it measure up?
12. Why does this (not) feel like the way ahead? Check your answers to the preceding questions. Is there something you didn't consider (enough)? Decide what to do.

Summary

This chapter has returned our focus to the 'bigger picture' to which the maintenance manager can make a very valuable contribution. Looking forward is something that can get neglected in the crush and immediacy of day-to-day imperatives. Buildings are important and valuable assets and they need to provide up-to-date facilities. This chapter has investigated questions of appropriate standards and how they may be determined.

The remaining chapters consider ways that improved facilities may be delivered.

References

Chartered Institution of Building Services Engineers and Building Services Research and Information Association (2007) *Refurbishment for Improved Energy Efficiency: an Overview*. CIBSE, London.

Geertz, C. (1973) *The Interpretation of Cultures*. Basic Books, New York.

Hastoe Housing Association (2004) *Good Practice Guide – Refurbishments*. HHA, Kingston upon Thames.

Hofstede, G. (1980) *Culture's Consequences: International Differences in Work-Related Values*. Sage Publications, Beverley Hills.

Kaplan, R.S. & Norton, D.P. (1992) The balanced scorecard: measures that drive performance. *Harvard Business Review*, Jan/Feb, 71–80.

Plimmer, F., Pottinger, G., Harris, S., Waters, M. & Pocock, Y. (2008) *New Build and Refurbishment in the Sustainable Communities Plan: Information Paper IP 2/08*. BRE Press, Watford.

Sustainable Development Commission (2006*) Stock Take – Delivering Improvements in Existing Housing*. SDC, London.

Trompenaars, F. & Hampden-Turner, C. (1997) *Riding the Waves of Culture: Understanding Cultural Diversity in Business* (2nd edn). Nicholas Brearley Publishing, London.

Yates, T. (2006a) *Refurbishing Victorian Housing: Guidance and Assessment Methods for Sustainable Refurbishment: Information Paper 9/06*. BRE Press, Watford.

Yates, T. (2006b) *Sustainable Refurbishment of Victorian Housing: Guidance, Assessment Method and Case Studies*. BRE Press, Watford.

17 The rehabilitation process

The previous chapter returned our focus to the 'bigger picture' after several chapters devoted to detailed consideration of defects. The maintenance manager can make a very valuable contribution to the longer-term and larger issues related to the future use of buildings. Questions of appropriate standards were addressed together with how they might be determined. This, the penultimate chapter, looks at ways that improved facilities may be delivered before the last chapter considers how the new and improved facilities may themselves be then kept up to date and up to scratch.

Preparation

The success of a rehabilitation project is determined in large part before it starts on site. It may not be overemphasising to suggest that the three most important requirements are preparation, preparation and preparation. There is much to be done.

Key elements here are to ensure that there is sufficient time for effective planning and that the right people are involved. An experienced group of 'change agents' may be able to work quickly and effectively to produce proposals for the change process. Less experienced people must be expected to take a little longer and maybe neglect some issues that turn out to be important in seeing through the change smoothly. Gains and losses incurred during the change process could be appreciable when set against the anticipated gains to accrue after the change, so it is important to get this right.

Briefing

Architects and surveyors may be good at design but they are not necessarily good at communicating with building users. A cynic may suggest that clients and users may be seen as an inconvenience whose main values are in paying the fees and in giving scale to the photographs. Let us plan for better than that. Sometimes the technical processes involved in developing and presenting the proposals are somewhat excluding of 'ordinary' or lay people, unfamiliar with the language of architectural representation. People are not always familiar with how to read a plan, to visualise a space in three dimensions from a two-dimensional drawing.

In some ways the existence of the building 'as is' is an advantage in giving three-dimensional form to the drawings, but it can also be an inhibiting factor in trying to visualise it newly reconfigured with new entrances, walls or doors added or removed, staircases moved, and so on. Drawings tend to show physical arrangements as 'existing' and 'proposed'; what we are concerned with here is not so much 'before' and 'after' as 'during'. There is a sense in which this is understandable; what is being 'sold' is the desired end-result. The scale of the upgrading is such as to make the intervening hardships seem worthwhile, and maybe not even thought about much at the time of seeking approval to the proposed improvement. Dealing with this inconvenient interlude in between, and making it bearable, is often very much an afterthought.

A smooth transition, maybe almost imperceptible, with minimal fuss and upheaval is a valid hope, and a realistic expectation if properly planned and executed. People want and need to be 'in the picture'. Involvement from the outset is important. On the whole people respond better when they know what is happening (and better still why), even if they don't like it, than if they are kept 'in the dark'. Of course this will have been happening if staff representatives have been involved in the development of the design proposals, and especially so if the organisation has that kind of consultative, maybe even participatory, expectation and practice. If however the organisation is rather more authoritarian, however benevolent, it will be doubly important to announce proposals in good time for those affected to at least come to terms with the proposed changes and how they will be impacted.

Time: the essence

Projects need programmes. How long is the building work going to take? How sure can you be? What will happen if any one task takes longer than expected and allowed for; what knock-on effects will there be on related tasks? How long should any particular task take; how does one work it out? It is not proposed to offer standard or target times here – there are many books related to construction project management. In broad terms, duration may be halved by applying twice the labour or doubled if operatives work only half the days planned. Egbu (1997, 1999) has identified the particular challenges of refurbishment management, and skills, knowledge and competencies required. The key with rehabilitation projects is to remember that existing buildings are not always as expected, so contingency times and money are needed; and there will be more complications still when the building is also occupied while the work is going on. Nutt *et al.* (1998) considers the management of risks of construction work in occupied premises, while Fawcett and Palmer (2004) offer guidance on refurbishing occupied buildings based on studies of a number of projects, from relatively simple repainting and recarpeting to more complex projects including structural alterations. Kelsey (2003) presents a study of issues related to the refurbishment of concourses and related retail space at major railway stations.

The need to move people around, and how to minimise the disruption, really needs to be considered at the design stage, or better still as part of the brief. Is having a large number of small moves better or worse than a 'big bang' one-off approach? Both approaches stand the risk of things not going according to plan and of consequential effects – they need to be thought through. This 'move design' takes time, which also needs to be built into the programme.

Regarding the duration of the construction works, it is as well to engage a contractor in discussions about what is realistic. (In this regard I include in construction all the works to and in the building to get from the before state to the desired end condition.) Issues related to fire risks and escape routes (Hinks & Puybaraud, 1999) and waste making, collection and disposal (Hardie *et al.*, 2007) will be important. Regarding timetable, it will be unhelpful to set unreasonable target dates for completion, or for that matter commencement, of the work and have to suffer more disruption as the works drag on leaving everyone unhappy. It is invariably a better feeling when works are finished early rather than late.

Place

While many aspects of design and planning could, at least in theory, be carried out anywhere – including these days anywhere in the world with electronic connection – some tasks are better done *on site*. For instance, people who are to have their jobs radically altered, or to be relocated, ought not to receive this unwelcome information in the form of an email. Similarly, building form and condition could be assessed remotely – the quality of electronic photographic transmission can be excellent – but this stands the risk of missing important detail; it also misses the use of other senses, which may be important. For instance noise from traffic or ventilation fans; kitchen or toilet smells – these could be important.

It is definitely worthwhile to spend a period of time on site communicating with *all* those likely to be affected by the projected works. If users' views are counselled at the outset they are more able to have an influence in the scheme that arises. More detail can be collected in face-to-face discussion than by electronic or paper-based questionnaires. Much depends on the questions asked, including how they are framed, whether they seek yes/no or open responses, whether there is scope for other matters to be raised. Thought also needs to be given to the relative virtues of company/department/team meetings and/or individual one-to-one discussions. Is it better to present a proposal for discussion/approval or to come with a completely open mind? Is there a particular kind of approach that the organisation prefers? Bear in mind that the more one seeks users' views the harder it may be to justify a solution that is out of line with broadly or strongly held views, especially if these have been voiced in open forum. People who have been invited to 'have their say' may expect to 'have their way'! Whatever proposals are presented, those presenting them must be able to justify them; they will also need to be seen to respect objections and alternative proposals. It helps to be able to say that such alternatives have been or will be considered. A cynic may say that 'sincerity is the key; if you can fake that you have got it made'. I advocate authenticity – people need to know they can trust you and what you say. So do not promise what cannot be delivered.

A visit to a similar project may help. Views from people involved in such projects can be very powerful, especially if delivered by them either in person or on camera rather than just quotes; they are doubly powerful if from people who had been sceptical and now see the benefits accruing from the new arrangements. Taking questions and answering them honestly and fully (including 'I don't know' or 'I'll find out') helps put people at ease and reduce potential 'us and them' attitudes.

Whether the design team is located on site may depend upon the size and scope of the project. There may not be room in the building – after all a main driver of the project is likely to be lack of space – so perhaps space in a building adjacent or close by may be appropriate. Whether the space is needed only for a short period may also be a factor. Perhaps a meeting room, or an empty flat, may be 'commandeered' for the duration. A temporary building or visiting caravan may be suitable. If there is no space on site available for the team perhaps a small meeting room in which a member of the team may be present periodically, e.g. every Monday, or lunchtimes, for a 'drop-in', or by appointment, may work well.

It will also be useful to have a presence on site while the works are in progress. If no physical presence can be facilitated, perhaps a notice board in a staff-room or a display in the foyer may be used to provide updates on progress with the project. Pictures of, or real-time views into the work in progress, may help. A webcam could be used to demonstrate on-line, although this could be dispiriting when watching paint dry! A point of contact will certainly be needed: a named person who can be telephoned or emailed, and an expectation that the issue will be responded to and dealt with in a prescribed time, just like with day-to-day maintenance requests. The personal touch is important to confirm that 'your call is important to us' – *you* are important.

Specifications

I said a little earlier about drawings, and how some people find them hard to read; I will add a little here about specifications. Building users may also find these difficult to understand; they are however a key indicator of the kind of standard that is being applied to the building in which they may be living or working. The user may not appreciate the difference in value between a 3 mm floor covering and one of 2 mm, but they may well understand the difference between vinyl sheet flooring, ceramic tiles and marble. So perhaps an indication, maybe a small display, of proposed finishes, furniture and fittings, ironmongery and so on may be appropriate. Perhaps choice could be offered. It may also be a good idea to give some kind of assessment of environmental performance – that one proposal or choice is particularly 'green', or 'un-green'. Beware though, people may not elect for your preferred option!

For newbuild construction there has been a movement in favour of performance specification, where the architect defines what it is that an element has to achieve by way of its performance, leaving the contractor to devise and decide how to attain it; for rehabilitation projects that is more difficult. Should one define the standard to be achieved in the building after its rehabilitation, or the amount by which the existing is to be improved? Either is possible and supportable. Certainly bearing in mind the cost of fuel to warm and cool the building it is worth looking for significant improvements in energy efficiency from the project; such expectations could be specified.

It is quite likely in a rehabilitation project that some existing features are to be retained and others, whether replacement or additional, to be provided to match. This can be difficult. The originals may be to an obsolete pattern; some may be broken during the works. It may be possible to track down some to match from another building, or to recover some from units that are to be scrapped or from a reclamation company. It may be possible to manufacture new to special order. The specification should make clear in advance what is acceptable so that planning and ordering can be done in good time.

Decanting

The decision whether to have the rehabilitation work carried out in an occupied building should be taken soberly, and the conclusion conveyed to those who will be affected as described above. Hopefully this decision will have been taken with at least a provisional programme of moves already prepared; if not this will need to be prepared soon. From a contractor's point of view, the greater the amount of vacant area at any time, and the longer it is continuously available, the better. For instance, while it may look attractive to a client to have the construction work carried out at weekends, this would involve building users packing their work away every Friday and taking it out to start again on Mondays. The building contractor could close off areas, cover up furniture, etc, clean and re-open, probably using subcontractors to do that work. The repetitive start-ups and close-downs will, however, be disruptive and costly.

Similarly, for staff to have to pack up, move, unpack, work, repack, move back and unpack again in the final place can be disruptive, and staff may be unsettled and less efficient for the whole of that period. In many ways it is better to be able to move direct from the old location to the new. This may be difficult, however, and undesirable in the case of housing rehabilitation, where on the whole people wish to return to 'home'.

Decanting needs to be carefully planned and executed. People's personal belongings must not be broken or lost in the move processes. Particular care will need to be taken of some possessions that, quite honestly, it will be difficult to understand why they are being kept, let alone valued. There are specialist firms that offer move management services. This experience may be particularly useful where specialised or heavy material is to be moved. It is not reasonable to expect 'normal' staff to undertake this kind of activity. It makes sense to involve those who will undertake the physical moving at the planning stage; they will have valuable inputs to offer. They will be able to estimate how long various matters may take and how many people will be needed to pack, to move and, if necessary, to unpack too. There are pros and cons as to whether organisations should have their own staff pack their own items.

Factors in favour of packing (and unpacking) your own items include:

- You know where you put things.
- If you forget you can only blame yourself.
- Similarly if you break something.
- You can pack at a time that suits you.

Factors in favour of having a specialist remover pack include:

- Skilled people pack more quickly and efficiently.
- They know how much can be carried in terms of bulk and weight.
- They have the requisite equipment.
- They supply strong and effective crates.
- Usually the crates can be supplied some time ahead and be left for sometime after the move if some staff prefer to 'do their own thing'.
- They carry appropriate insurance.
- They can advise.

Needless to say, perhaps, but a plan is needed of who and what is going to be moved where, by whom, when and how. (We have already discussed the need to tell people why!) Such a plan can be quite simple or it may be complex. There could be quite a lot of interactions – that person A1 cannot move to position 8X until person B4 has been moved out into position 3N. These plans need 'buy in' from all the participants (or at least as far as that is possible – it will be difficult to please everyone) before putting them into operation. The move manager also needs to be around at the time of the move to help sort out any issues arising. Even the best laid plans are likely to incur some problems along the way: 'The best laid schemes o' mice an' men gang oft a-gley' (Burns, 1786).

The new locations for people, however temporary, need to be ready in time to receive their new occupants. They should be checked before the move. Issues likely to need to be faced are such as:

- telephones in place and/or new extension number/direct line allocated through switchboard, so that user has continuity of contact number;
- similarly fax provision;
- security access permitted to new areas; codes changed; denied to old areas;
- notifications of pending moves and moves made;
- welcome arrangements – flowers, coffee, etc.;
- recognition of potential for staff upset (could be an emotional time, especially if established teams are being reorganised);
- need for help.

Consider as far as possible what could go wrong and what will be done in that event. It may be unlikely that it will go wrong, but you will sleep better the night before if you know you have contingency plans in place.

If several moves are planned, start with a relatively small move as a pilot and allow time between that move and the next in which to review and revise plans. If you leave the biggest move until last you can also feel (and maybe celebrate) when that's completed. The move manager should make it his or her business and priority to be seen around and to be asking 'how's it going?'.

Ask those who have been moved to complete a short feedback form from which you can learn what went well and not so well from the point of view of the 'real' client, those actually moved. These 'displaced persons' should also be checked up on regularly while they are in their temporary locations. They may be expected to suffer some of the symptoms of refugees, missing their old 'home' and friends, unsure how they will adjust, fearful of the future. 'News from home', including assurance that all is well and on programme for the next stage and perhaps occasional visits to see progress may be particularly valuable in maintaining morale. Communication will be as important as ever.

Work in progress

While staff are in their 'temporary homes', work will hopefully be proceeding apace on the spaces vacated. If not, they will understandably want to know why not. It may make sense to leave a little contingency between the space being vacated and builders moving in (just in case of delays in vacating), but this should not be too long or dissatisfactions will be voiced. Arrangements should be made for periodic progress reports to be made; it helps to highlight when milestones are reached. It will be good to announce, for example, 'The new kitchens are now complete (see photo); the new electrics are well on the way and we are on schedule (in fact marginally ahead) for completion of all the work ready to move back in as planned on . . . '. It is good to report good progress; it is vital to report when work is delayed so that there are no surprises if and when moving back is delayed.

Work in progress on a rehabilitation project should be monitored and managed at least as closely as any newbuild project, almost certainly more so. The need to be able to agree changes in the light of matters arising is a semi-constant. There are bound to be changes. Walls will be found to be bearing unexpected loads; floors joists will have rotten ends; deleterious materials will be discovered. All these cannot just be covered up and left; they have to be considered and dealt with. Building users don't need to be informed about the detail of these things other than when progress is affected, but someone needs to be present who will represent the client's best interests.

On a construction project there will normally be regular site meetings at which progress is reported and recorded, discussions had about problems, and minutes made including a record of actions decided. Typically on a newbuild project formal site meetings may be held monthly. For a rehabilitation project weekly meetings are more likely, and it may be valuable at the beginning and toward the end to have daily updates. These will enable plans for occupation, and in the case of decanted personnel re-occupation, plans to be made, and perhaps advanced or delayed.

Most changes needing to be made to rehabilitation plans will be minor, but their cumulative effect is what tends to cause delay. A comparatively small discovery may mean that a piece of work needs to be postponed until after a little redesign is done. This may entail, for instance, standing down the plasterers who were lined up to continue their work there; they may go off to another job *pro tem* and it may be some time before they are again available. They need to work to earn and will not like being 'messed around'. Subsequent work may be delayed unless it is reprogrammed and more labour applied. The consequential effects on the erstwhile occupants of the space may be considerable – they may need to vacate the space into which they were decanted!

It may be possible to seek to reclaim consequential losses from the building contractor but that is likely to be unfruitful. If the contractor is to be held liable under the terms of the contract he or she will certainly price the work accordingly. He/she will also resist such a claim by recourse to contract terms which will not hold him or her responsible for the effect of extra work required that was not specified at the outset. Indeed the contractor will normally expect not only an extension of time on the date for completion but also an additional sum in payment for the additional work and delay.

It is common to seek to solve construction-related problems as they arise so as not to delay works unnecessarily and to leave sorting out the financial and other consequences until after practical completion of the works. This may be understandable but it cannot be considered good practice. Consequences should be considered at the time that they are reasonably foreseeable. The effects on time and cost should be calculated and recorded at the time and preferably agreed. No client enjoys extra costs but contractors are entitled and expect to be paid for extra work. They are also entitled to profit on work they expected to undertake but was cancelled at a late date, and to be paid for work entailed in rearranging labour, and returning and reordering materials even if the revised requirement is for less than previously needed. Change, even reduction, costs money.

If work is over-running, it is worth exploring what could be done to still complete works by the due date, so that the plans for moving back do not need to be remade. Depending upon the reasons for the over-run and who is liable, it may be that the contractor should be expected to do whatever is necessary to remedy the situation so as still to complete on time, all at his own expense. Even where the delay is not the contractor's fault he will still be expected to use his best endeavours to retrieve the situation. The client should also recognise, bearing in mind the costs and inconvenience of staying longer in the temporary accommodation – if that is even possible – that it may be in his or her best interest to pay more for an earlier completion. Delays, however well they may be explained, are 'bad news', leading to disillusion, irritation, frustration and possibly disaffection; they may be used by those who were unenthusiastic about the project as a vindication of their position, leading to further grumbling. It is good to get on top of issues and deal with them.

Preparing to move back

It may be thought that moving back will be much the same as moving out was. That may be an over-simplification. Being decanted was recognisably a temporary move; but moving back is a permanent move so expectations will be higher. People will have hopes that all their previous problems will have been overcome. That is a tall order and probably unrealistic, but still some will have that hope. It is important, therefore, to do what can be done to ensure that the work is complete.

It is as well not to agree a date for moving back until *certain* that all will be well, all complete before the day of moving back; no wet paint! An alternative approach might be to agree a firm date on the basis that all unsatisfactory work will be corrected by an agreed date – that is very much an undesirable situation; the subsequent disruption will be very irritating to those who have been looking forward so much to being back. There will have been enough difficulties to get used to in the temporary location, and almost certainly some in the moving back process to deal with, without continuing hassles. Don't move back to problems. Don't accept building, decorating or fitting out works as complete if they are not (see Chapters 8 and 9).

Check

A checklist should be developed that reflects all the important things that need to be in place for the move back to be a success. The list here should be considered as no more than a draft to prompt the development of an appropriately specific checklist.

The numbering here is of no particular significance other than for future reference and to indicate something of the potential scale of the exercise.

1. ALL structural work complete.
2. All fire safety in place: escape routes; signs; alarms; assembly points; sprinklers; hydrants; extinguishers, blankets, etc.; wardens.
3. All services (re)connected, tested and FULLY functional; especially NO LEAKS!
4. Evidence of meeting all performance requirements handed over, e.g thermal and sound insulation; air-tightness.
5. All contractors' equipment and waste materials removed.
6. Floor finishes in place and movement paths free of obstruction.
7. All decorations complete, including making good; no defects to be corrected later.
8. All new equipment in place, connected and functioning correctly.
9. Operating manuals handed over, including contact details; spares received.
10. Service agreements?
11. All doors, windows, etc. intended to be openable, tested and satisfactory.
12. Defects liability periods/dates agreed; inspection and making good arrangements.
13. Access (including for disabled people), parking, transport arrangements in place.
14. Client's insurances in place.
15. Client's support systems in place and ready to swing into action on agreed date(s) – e.g. office cleaning, catering services, security, waste collection, etc. There is a lot to consider in this area.
16. Sufficient time available for users to prepare to move back.
17. Users fully briefed on how to occupy and use the 'new' building.
18. Key contacts known, available and with authority to deal with all problems during move back.
19. Move back plan agreed and published; to include who goes where and when.
20. Contingency plans.
21. Trial walk-through.
22. All requisite promotional activity in place, including to let customers know.
23. Notify adjoining owners/occupiers.
24. Hand back arrangements in place for the temporary accommodation, after everything is OK with the 'new'.
25. Plans in place for follow-up after the move to check all is satisfactory.
26. Other?!
27. What have we forgotten?

Post-occupancy evaluation (POE) is considered in more detail in the next chapter.

Summary

This chapter has looked at ways that improved facilities may be delivered through a rehabilitation project. It has considered the value of careful planning of operations, including the need for operations to continue smoothly whether in the existing building or by decanting. The moves around, out of and back into the building also need to be prepared with care, and with appropriate contingency planning. There is much that can cause delay and frustration in a rehabilitation project.

The following chapter, the last in the book, reminds us that the new and improved facilities will themselves also need to be kept up to date and up to scratch.

References

Burns, R. (1786) *To a Mouse.* In: *Oxford Dictionary of Quotations* (4th edn) (1992). Oxford Universtiy Press, Oxford, p. 163.

Egbu, C. (1997) Refurbishment management: challenges and opportunities. *Building Research and Information,* **25**(6), 338–347.

Egbu, C. (1999) Skills, knowledge and competencies for managing construction refurbishment works. *Construction Management and Economics,* **17**(1), 29–43.

Fawcett, W. & Palmer, J. (2004) *Good Practice Guidance for Refurbishing Occupied Buildings.* Construction Industry Research and Information Association, London.

Hardie, M., Khan, S., O'Donnell, A. & Miller, G. (2007) The efficacy of waste management plans in Australian commercial construction refurbishment projects. *Australasian Journal of Construction Economics and Building,* **7**(2), 22–28.

Hinks, J. & Puybaraud, M.C. (1999) Facilities management and fire safety during alterations, changes-in-use and the maintenance of building facilities – a management model for debate. *Facilities,* **17**(9/10), 377–391.

Kelsey, J.M. (2003) *Drawing the line: balancing the spatial requirements of customer and contractor in occupied refurbishment.* Proceedings of Joint Symposium of Conseil International du Batiment et de Recherche Working Commissions W55, W65 and W107, 'Knowledge management', 21–24 October, National University of Singapore.

Nutt, B., McLennan, P. & Walters, R. (1998) *Refurbishment of Occupied Premises: Management of Risk under the CDM Regulations.* Thomas Telford, London.

18 New life in the building

The previous chapter looked at how improved facilities may be delivered through a rehabilitation project. Careful planning of operations, including the related moves around, out of and back into the building, was seen to be important, together with the need for appropriate contingency planning.

In this, the final chapter, we are reminded that the new and improved facilities will themselves also need to be kept up to date and up to scratch. Techniques and timings are considered together with some thoughts about motivation and moving on.

How was it for you?

Having carried out maintenance or improvement work it is important to evaluate how well it went. Not only is this a vital activity from which to learn and to improve how things are done next time, but it also demonstrates a genuine interest in the people who have been involved and affected, as well as the results.

There will almost certainly be a next time. It may be soon; it may not be for some time – it will probably differ, perhaps in major ways. Personnel involved will almost certainly be different, and even if some are the same people they too will have learnt and been changed by the experiences gone through then and since. Even if things went perfectly last time unfortunately that is no guarantee of similar success another time; it may, however, be a good starting point.

Problems, shortcomings, errors, mistakes, things which 'didn't go so well' – these are all learning opportunities. They are not to be dismissed, forgotten about or glossed over – they should be learnt from. Similarly, the people who made those mistakes ought not to be dismissed, or at least not without full discussion and consideration. There are in any case, as is right and proper in an enlightened world where people are valued and respected, proper procedures to be followed in relation to discipline and dismissal of personnel. Ask yourself, would you like to be sacked for a mistake, perhaps a genuine misunderstanding or miscalculation? Such experiences can be chastening. Employment law is beyond the scope of this book. The best advice I can give here is to avoid getting into legal disputes, and, when you do, seek professional advice. Legal advice will cost you money of course, but so will going into court, especially if you lose.

There is an apocryphal story of an employee who makes a horrendous error, losing his organisation more than a million pounds or dollars or euros. The employer elects not to sack him or her on the basis that they have now invested over a million pounds in the individual's development, from which they hope to derive benefit over the years ahead and not to give a competitor the advantage of that learning.

The learning organisation

All individuals and organisations are capable of learning from experience. Indeed Argyris and Schon (1978) argued that people learn better from experience in the workplace than in the classroom. Revans (1980) suggested that ideas should be tried out and evaluated in practice – a process he dubbed 'action learning'. This was further developed by Kolb (1984) into what he called 'experiential learning', applying Deming's plan–do–check–act cycle. Every action is reviewed afterwards to inform the next cycle of action. Senge (1990) brings all these together in what he calls a 'learning organisation'. Such an organisation seeks to learn from every opportunity and to formalise that learning (Handy, 1989).

Armstrong (1994, p.76) defines a learning organisations as 'an organisation which facilitates the learning of all its members and continually transforms itself'. He summarises the features of a learning organisation as identified by Mumford (1989); these include:

- Staff identify learning needs and set themselves learning goals.
- Managers aim to provide learning opportunities.
- Training is provided.
- There are regular feedback and review.
- Mistakes are tolerated as long as they are learnt from.
- Managers' performance in helping staff develop is reviewed.
- Established ways of working are challenged.

If this kind of approach to formalising and 'capturing' learning is followed consistently it provides opportunity for continuous improvement. After all, learning can only really be said to have taken place if it informs and has effect upon future actions, hopefully for the better. If processes are changed in the light of experience, but the hoped for improvement does not in fact occur, then further learning has taken place, which can be taken forward in the form of further or different changes next time.

Japanese industry of the 1980s improved its quality and performance significantly through applying *kaizen* (*kai* means change and *zen* good, or better), a process of continuous improvement.

The focus is a forward-looking one; one looks back in order to improve the future.

Figure 18.1 illustrates how experience may be used to inform future decision making. The diagram takes the form of a fountain, where information is collected at all stages of a project (design, commissioning and use) and can be applied at any stage. The earlier that consideration is made of such knowledge and experience the more impact it is able to have on the outcome in terms of better and more effective operation and use of the building.

Techniques and timings

Although the feedback and review processes may be seen as part of a continuous learning process, certain activities or stages may be more to the fore at certain times than others. Some processes may follow regular cycles, while others may be more irregular; some cycles may be short, some activities infrequent. It may be possible to plan some review processes on a cyclical basis; others may need to be 'as and when required'.

For example, some processes will tie in with and inform annual demands, such as preparation of budgets for the following financial year. Different, more strategic, reviews may be appropriate

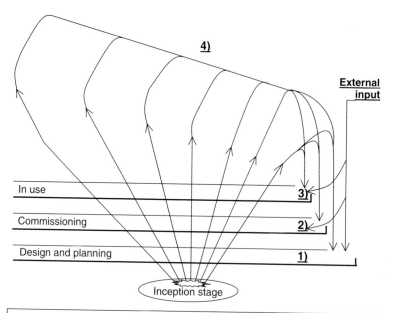

Figure 18.1 Learning from and applying experience; the fountain of knowledge.

periodically to survey the larger landscape and plan further ahead. Smaller but more detailed exercises may be targeted at specific issues.

This threefold approach is common and has much to commend it: periodic overview; annual review; occasional formal feedback.

Periodic overview

This is the longer-term, strategic, exercise – trying to look ahead 5, 7 or 10 years. It may take some time to complete, perhaps several months. It will typically be informed by previous annual reviews – recurrent items, enduring issues, trends (both good and adverse), etc. – and some 'crystal ball gazing', trying to forecast the future.

The questions to be asked will depend very much upon the kind of organisation and the context in which it is operating. The following may provide a starting point for the construction of various data gathering, opinion seeking and market testing exercises and related analyses; it could become a draft list of contents for the strategic plan, the production of which it will inform.

- Items from review of last (three) annual review reports, e.g.:
  - ☐ recurrent items;
  - ☐ matters that are getting progressively worse;
  - ☐ areas of poor sales; low client satisfaction.
- Forward and 'worse case' projections 5 years ahead of:
  - ☐ 'sales';
  - ☐ costs;
  - ☐ profits.

    These will help support the need for action upon poor performance.
- Buildings:
  - ☐ trends in operating costs, including energy, repairs, upgradings (disaggregated);
  - ☐ condition;
  - ☐ 'fit' between buildings and current needs;
  - ☐ opportunities for active re-use.
- 'Competitors'/similar organisations:
  - ☐ What are they doing more of?
  - ☐ What are they doing less of?
  - ☐ What are they doing better than us?
  - ☐ What are they doing new?
  - ☐ How are they perceived in the 'market'?
- Three ideas to work up/assess feasibility:
  - ☐ How easy to implement; how soon; what cost?
  - ☐ Potential downsides?
  - ☐ Risk from going that way; of not?
  - ☐ Alternative(s)?
- Internal organisation:
  - ☐ How adaptable? How well has 'organisation' adapted?
  - ☐ Current shortcomings; 'empires'?
  - ☐ Scope for minor adjustment of roles, responsibilities?
  - ☐ Need for radical change?
- 'Systems':
  - ☐ reporting: topics, regularity; what happens?
  - ☐ relevance;
  - ☐ value added;
  - ☐ communication generally.
- Abiding 'values':
  - ☐ How do we see ourselves?
  - ☐ How are we perceived by others?
  - ☐ Corporate social responsibility (CSR)?
  - ☐ Environmental performance; waste, 'greenness'?
  - ☐ Ethics; fair trade; business practices?
  - ☐ Respect for people; trust; working conditions?

- Social:
  - □ How 'good' an employer?
  - □ Workplace stress? Healthy workforce?
  - □ 'Benefits'?
  - □ Work–life balance?
  - □ Recruitment and retention?

Each area can be assessed in terms of Kipling's six friends – who, what, where, when, how and why – as prompts to further or underlying issues. It is more important to have an enquiring mind and to (as a building pathologist or surveyor would) 'follow the trail' than to complete a checklist. This kind of strategic review happens infrequently. It provides a rare opportunity to stand back and think – even to 'think the unthinkable'. This is the time to be radical in thought, whether or not that is seen through into action. Decisions taken at this stage may set a course for future success, or survival, or not. For this latter possibility it is doubly important to review regularly how well the organisation is performing against the plans it sets. Quarterly, monthly, or even weekly review may be needed to see that all is well, and will form a basis for the annual review process.

Annual review

An annual review is likely to have a strong reporting focus. A lot of its content can be assembled and aggregated over the year. The topics on which to report should be a composite of standard items of interest to any organisation and issues highlighted as significant in the organisation's strategic plan derived from the major periodic review.

It is largely aimed at establishing and demonstrating that the organisation is on target and meeting its various goals. It often, therefore, has something of a backward-facing focus, recording what has been done and has been achieved. It may have some forward projection, largely in terms of proposed adjustments (up or down) in the light of the reported progress (or lack of it).

Statistics relating to performance are a significant element of any annual review. There will often be numerous tables of data (sometimes so many as to make it hard to 'see the wood for the trees'). These tables will often have several columns of figures recording such as:

1. Planned output.
2. Actual output.
3. Difference: planned minus actual (brackets = shortfall).
4. Planned input.
5. Actual input.
6. Difference: column 4 minus column 5.
7. 'Efficiency': column 2 divided by column 5.
8. Strategic plan target for coming year.
9. Proposed (revised) target.

It will be necessary somewhere to define how each item is to be measured in order to ensure consistency across the organisation and across the years.

The columns indicated above may be 'enhanced' by comparisons (perhaps by showing figures in brackets, or in *italics* or **bold** type) with figures from last year. There may also be a need to provide a brief note to explain the odd or unusual figure, perhaps where one item relates only to 9 months of the year due to some major change or significant influence – maybe an element 'out of service' for a period.

All significant variances should be commented upon with explanations given for their occurrence. For maximum benefit there should also be recommendations for action(s) proposed to be taken.

An assessment of performance of each significant element and of the service overall should be made in the conclusions, with particular attention being given to the elements that performed especially well or badly. Each of these should give rise to a recommendation (or several recommendations) for action, together with an indication of its urgency. It is particularly helpful if all these recommendations can be assembled into a single prioritised list, and better still if accompanied by estimates of cost.

Typical items for report might be such as:

- number of call-outs;
- mean response time;
- total maintenance expenditure;
- breakdown of expenditure per element/per building/per month.

Possible headings for an annual review might include for instance:

- Summary:
  - □ How are things overall?
  - □ Be consistent and accurate in your use of 'evaluative' terms, such as: excellent, outstanding, good, very good, fair, poor, bad; those which compare against some standard, like acceptable, sufficient and unsatisfactory; and those which compare with a performance previous or elsewhere such as better, worse, significantly worse, improved, higher, lower, superior.
  - □ Keep it brief (500 words maximum).
- Highlights of the year:
  - □ Include photographs especially of people and of new or improved buildings, preferably combining people *and* buildings.
  - □ Start with the most significant event/improvement of the year.
  - □ Other significant happenings: personal achievements, awards, long-service.
  - □ Arrivals and departures (especially retirements).
  - □ Births, marriages, deaths.
  - □ We look forward to . . . in the coming year.
- Performance over the year:
  - □ Explain rationale for this presentation, e.g. by building/service/division.
  - □ Include tables, charts and graphs of various kinds (e.g. lines, bars, stacks, pies), ensuring the presentation is appropriate (e.g. no pie chart showing 50% this and 50% the other!).
  - □ Comment on all significant differences: between sites; over time.
  - □ Avoid odious comparisons (any bad news should be communicated separately and preferably face to face); the annual report should be truthful but not brutal.
- Particular projects (may be several chapters depending upon number, range and significance):
  - □ Brief descriptions of each, including some measure of size or scope, e.g. n square metres, *x* pounds, our largest yet, the first phase of our five-year plan to . . .
  - □ Benefits accruing already (or expected) (or even better than expected).
  - □ Focus on some of the team involved; include quotes.
- Main points from the year, including areas for attention:

    ☐ Major service improvements that came 'on stream' during the year, and benefits already accruing.

    ☐ Changes made which perhaps didn't achieve what was expected, understandings of why that might be, and what has been or is planned to be done as a result.

    ☐ Schemes in preparation and their intended benefits.

■ Looking forward:

    ☐ Priorities for the coming year.

    ☐ Continuations of programmes already in progress.

    ☐ New programmes and projects.

    ☐ Indications in broad terms of whether and how much these are increasing (or decreasing) and why.

    ☐ Expected highlights of the year ahead.

■ A reminder of our longer-term plans and intentions:

    ☐ How these fit in with and further the organisation's values, aims and current programmes.

    ☐ They are an indication (a measure?) of our continuing/increasing commitment.

    ☐ Caveats where necessary, e.g. where dependent upon funding sources.

■ For further information, contact .... (person; address, phone, fax, text, email, etc.); also available in large print, Braille, range of languages, on website; summary document available for friends, clients, etc.

■ Appendices, including financial summary.

The report must be produced with the intended audience in mind. It is worth bearing in mind that persons other than those expected might read the document too. Therefore it is important to consider issues related to confidentiality – competitors will be interested to know of your plans, and difficulties, so some circumspection may be appropriate. Copyright may also be an issue; it is not, for example, acceptable to use photographs taken by others without permission and attribution.

It may be appropriate to produce variants of the report, for instance a confidential 'warts-and-all' version for consideration at board level, perhaps with subsets of material for discussion in other forums; and a more glossy version for publication, distribution to friends and media, etc. What is important is that the document, and what is conveyed therein, has been *informed* by the data collected for the review.

Formal feedback

While such exercises may be conducted at irregular intervals, they will often be prompted by particular projects or events. It may be possible on occasion to work to a standard format, which would enable opinions or performance data to be compared between projects or over time. However there will almost certainly be many issues to be probed that are of specific or especial interest in relation to the particular project or programme. Perhaps a pro forma or format for interviews may be constructed that combines standard and non-standard questions or headings. A blend of open and closed questions eliciting scaled responses (that can be analysed statistically) and personal views can give a rich expression of how well matters went.

When should such feedback be elicited? Examples may be:

■ *During a project,* for instance an improvement or rehabilitation project, it may be thought 'caring' to enquire of those affected:

    ☐ how much are they being affected – more or less than they expected – and in what regards;

    □ how well is the contractor and/or the organisation responding to complaints;
    □ degrees of optimism/pessimism;
    □ data regarding downtimes, sickness, staff leaving, etc.
■ *On completion* of a project, opinions could be sought on its success or otherwise:
    □ quality of information provided before commencement;
    □ satisfaction with consultation processes, and results thereof;
    □ quality of periodic updates on progress, delays, consequences, mitigation;
    □ satisfaction with end-product;
    □ perceived benefits accruing.

It may be useful to undertake a pre-project commencement exercise to ascertain the concerns of people who believe they will be affected (for better or worse, and by how much). Feedback results collected during and after the project can then be compared to see how well or not those and other concerns have been addressed.

It may also be useful to collect opinions and data a few months after completion of the project, when perhaps a 'steady state' has been achieved with immediate issues dealt with and hopefully overcome. Again results from this survey can be compared with those collected earlier. This may show how people's agitation reduces and memories dull over time, as people get used to the new *status quo* and accept the situation together with its previously perceived shortcomings.

None of these exercises is worth carrying out, however, if nothing will be done with the data. There has to be a culture in the organisation of genuine interest in people and the performance of the buildings they occupy. If nothing is done as a result of these exercises people will become cynical. They will then either withdraw support from further exercises, or give untrue answers to 'sabotage' the results, and maybe lose motivation in their jobs. Similar effects may be seen if there are too many questionnaires. Some people will feel that 'management' should already know what is going on, or should come and find out directly, or that they should 'get on and manage'. Perhaps some exercises could be carried out on a sample basis; and some may be made available by email or on the web for those that wish to respond. As long as concerns are addressed about potential bias in 'self-selecting' samples, these can still be very useful.

Post-occupancy evaluation (POE)

It is becoming much more commonly accepted that a new, or newly refurbished, building should be assessed after its completion to see how well it is performing. Interest was sparked by the 'Probe' series of post-occupancy review studies of Leaman and Bordass published in *Building Services*, the CIBSE Journal (Leaman & Bordass, 1995). The studies showed that many of the buildings investigated did not function as well as had been intended. The Probe studies included a range of components, such as:

■ review of design intentions;
■ design and construction documentation;
■ walkthrough observations;
■ energy surveys;
■ occupant questionnaires;
■ management interviews;
■ designer responses.

Resultant recommendations included:

- avoid complexity;
- monitor performance;
- make feedback routine.

Leaman and Bordass have followed up this work with studies related to workplace productivity (Leaman & Bordass, 2000), and their POE methodologies can be found on the website of the Usable Building Trust (www.usablebuildings.co.uk). Bordass *et al.* (2001) produced a summary of conclusions and implications of the 'Probe' post-occupancy studies, and a portfolio of feedback techniques has also been published (Bordass & Leaman, 2005). In the USA, Wolfgang Preiser has been a pioneer (Preiser *et al.*, 1988; Preiser & Vischer, 2005).

Matters are getting more formalised and consistent in UK. The Building Research Establishment (BRE) was commissioned by the then Department of Trade and Industry to report on POE (Jaunzens *et al.*, 2003). Survey methods can be found on the web at www.projects.bre.co.uk/earlypoe/.

The higher education sector seems to have become particularly interested in POE. The Higher Education Funding Council for England (HEFCE) jointly funded a project with the Association of University Directors of Estates (AUDE) (HEFCE, 2006), and the Scottish Funding Council (SFC) issued guidance in 2007 (SFC, 2007). The HEFCE document surveys the POE 'landscape'; it advances no one definition of POE. It identifies that POE may be engaged at different levels: an operational review (at 3–6 months after hand-over) and a performance review after 12–18 months. Templates are included, e.g. a four-page pro forma per room for an observation evaluation sheet. Advantages and disadvantages of a range of techniques are reviewed, including:

- walkthrough;
- interviews;
- focus groups (of six to eight people; six questions maximum);
- workshops (half-day);
- questionnaires (20 minutes maximum);
- measurement.

The HEFCE document offers a good overview of POE including material related to its value, and with the advantage of ready-made templates to work from. Personally I felt there was a lot of data to be collected and I was not sure of the long-term worth of that amount of data. However, the ability to work to a common format across the sector and compare results across institutions could be a definite asset if data is recorded consistently.

A guide has been published by the British Council of Offices (Oseland, 2007), and there are several organisations offering POE services. Buro Happold, a highly rated firm of engineers, states on its website:

> *During a 50 year life cycle, saving just 1% on energy costs would pay for a post occupancy evaluation and monitoring analysis.*

Pros and cons: why POE?

The principal motivation for a POE is most likely to be a 'fine-tuning' of the building so that it operates as effectively as it can. It is likely that there will be a number of aspects where things have not turned out quite as well as hoped or expected; and there may be some where early attention might be warranted.

POE provides opportunity for the organisation to demonstrate that 'the management' are genuinely interested in:

- how well their buildings are functioning;
- how well projects work out;
- how is it going for the staff.

It also enables whoever suggests the POE exercise to show his or her line manager that they have these interests at the heart of their concerns. They want to do a good job; and be seen to be doing so too.

The Hawthorne experiments of the 1930s (Mayo, 1933, 1949) identified what has become known as the 'Hawthorne effect', defined as 'a short-term effect caused by observing worker performance' (Landsberger, 1958). In the studies on illumination, a set of selected workers was submitted to a variety of lighting conditions. The investigators found that even when the lighting levels were reduced the workers' output increased; even when no change was made (the investigators having only *appeared* to change the lamps), performance increased (Roethlisberger & Dickson, 1939; Veitch, 2000). So the carrying out of a POE may of itself increase productivity, at least for a period.

A POE exercise is, of course, a learning opportunity too. It provides learning about the matters above and more:

- management interest;
- how well the buildings are functioning;
- how well the project worked out;
- staff concerns;
- how to carry out a POE;
- how to further improve the building;
- how to carry out a rehabilitation project better next time;
- how to carry out a POE better next time.

All activities need to be justifiable. The potential benefits to be derived from a POE exercise may be self-evident; they have been articulated above. However, that may not be enough. If the costs of a POE are seen as an extra, they may not be sanctioned; they need to be budgeted for as part of the project. The cost may appear to be quite high, especially as the project will have been expected to have addressed and rectified any pre-existing problems. Why spend extra money to ascertain that the project has achieved what has already been paid for?

Who should do POE?

Perhaps a way ahead would be for the contractor to commission a POE to demonstrate the effectiveness of the finished project. This could, however, have the effect of heightening differences of opinion about what makes for 'effectiveness' or 'success', and seeking to lay blame elsewhere, for example in terms of the project brief or design in which the contractor had no involvement. Perhaps the contractor might have a vested interest in putting a certain slant on questions – what is asked or measured, how, when, of whom, etc. – or 'gloss' on its interpretation or presentation. With a PFI (Private Finance Initiative) project, where the contractor has a continuing involvement with and responsibility for the effective operation of the building for maybe 20 or 30 years hence, a POE may very well be in their own long-term interest.

A POE exercise could be commissioned by the senior management of the organisation occupying the building or the landlord, depending perhaps on who is paying the bills, either for the rehabilitation or for the operation of the rehabilitated building. It could be commissioned by the maintenance or facilities manager or by the human resources (HR) department; it may be commissioned by the contractor or by the designer (architect, surveyor or engineer).

The exercise could be designed and/or carried out by any of the above in-house or by consultants. There are specialist consultants that can advise on, organise and execute POE exercises, and there are plenty of employment agencies happy to arrange 'temps' (temporary workers) to carry out the fieldwork if necessary. The decision on how best to take this forward will depend on a range of factors, such as:

- likelihood of future POE exercises;
- availability of suitable staff;
- in-house knowledge and capability;
- confidence;
- size of project;
- previous experience.

There is value in engaging consultants. They have the appearance of independence. I say 'appearance' because there is a sense in which 'he who pays the piper calls the tune'; there can be a tendency to 'bury', omit or rephrase bad news. Sometimes this censorship is self-imposed and unspoken, seeking to avoid embarrassment. This may be understandable but is unhelpful; better not to carry out a POE if it is not going to help identify issues and lead to improvement. If only good news is sought, then public relations (PR) people should be engaged.

Monitoring and refreshing

Having completed the building rehabilitation project, including moving out, moving around and moving back in, and the design, planning and management of all that, and having completed a POE, it is important to maintain the gains achieved, including improved motivation and production as well as building performance. The maintenance manager will have a key role to play in this.

In the ideal world, the maintenance manager will have been involved in the making of decisions at the briefing and design stages about layouts, materials, finishes, services installations, access, cleaning, maintenance requirements etc. He or she may have been involved in the construction stage – there will be decisions that were taken then that will have abiding effects through the operational period ahead. If not involved then, the maintenance manager should do what he or she can to talk to personnel from those earlier stages while they are still around. This will enable vital and missing information to be obtained and recorded for the future.

Fewings (2009) stresses the importance of taking a holistic view that sustains trust and transparency, while Hoxley and Rowsell (2003) have suggested the use of video recordings for knowledge transfer and learning. Pertinent parts of the construction process could be recorded to inform future maintenance purposes where the maintenance manager could not be present.

Documentation

There is no point in making records and creating voluminous files just for the sake of it – they must be useful. I am reiterating this point. However, there are some items which, if not attended to at the outset, will be difficult to deal with at a later date. At hand-over it is important to collect documents related to just what it is that has been taken over, in case they may need to be referred to (while perhaps hoping that will not need to be referred to) at some stage in the future.

Examples of documentation that should be held include:

▪ a full set of 'as built' drawings and specifications;
▪ contact details for contractor, including principal personnel;
▪ contact details for all suppliers, subcontractors;
▪ maintenance and operating manuals for each piece of equipment;
▪ guarantees and warranties;
▪ cleaning recommendations;
▪ as much information as can be gleaned about anything particularly unusual.

This documentation, together with the results from the POE exercise (and experience), should enable schedules to be drawn up for periodic and one-off maintenance activities for the year(s) ahead. This can be expected to include daily, weekly, monthly, quarterly and annual tasks and should, as far as possible, be balanced over the day, week and year. Some tasks will be more easily accomplished at certain times of day or week or year.

Activities likely to feature over the year might include:

▪ toilet and washroom inspections (hourly?), cleaning and replenishing (daily?);
▪ floor cleaning (daily);
▪ waste paper bin emptying (review whether this can be replaced by recycling);
▪ window cleaning;
▪ machine maintenance – lubrication, cleaning, servicing (check agreements);
▪ testing of electrical appliances; wiring systems; gas installations; *Legionella*;
▪ checking of asbestos and other deleterious materials;
▪ assessment of energy use; opportunities to reduce; item defects inspections, surveys, assessments of performance for annual review;
▪ planning, monitoring, reviewing;
▪ staff and management meetings; staff development and training;
▪ social events; 'team maintenance'.

Refreshing the building

It is worth reminding ourselves that as soon as the rehabilitation project is complete it will be deteriorating. Chapter 7 discussed this; and hopefully the advice in Chapter 8 will have helped reduce the scope for deterioration in the 'new' building by selection of more durable materials, etc. Even so, materials will deteriorate; systems will age and become obsolete; layouts will become unsatisfactory, and so on. The building will need periodic if not perpetual refreshing.

Hopefully the regular maintenance and review activities already discussed will mean that needs for further improvements will be routinely identified and incorporated into future plans as they are developed from year to year. The processes and attitudes within the organisation, should be such that proposals are welcomed, considered and put into action at appropriate stages of the various cycles.

Refreshing the organisation and staff

Buildings affect people. Buildings affect performance. It is therefore imperative to keep the building up to date and functioning well. On the whole, people will appreciate seeing investment being made in the buildings in which they live and work. They will not, however, enjoy being disrupted repeatedly, so it will be important to plan developments well, including consultations with those who will be affected. There will be less resistance if proposals are in line with people's own wishes, so there is much to be said for soliciting ideas from the 'shop floor'. This 'bottom up' approach (rather than, or together with, the traditional more autocratic 'top down') has, in any case, much to commend it; it testifies to the value placed on people.

There may well have been liaison committees and the like set up during the rehabilitation process; if so, it will make sense for this to continue. Not only will members have useful contributions to make in terms of feedback on the project processes, but it is likely they will also have picked up ideas for the future along the way. These representatives and those who found their involvement useful will also quite likely want arrangements to continue. Indeed to 'close down' on such arrangements may suggest to staff that their comments and cooperation are no longer needed or valued. Some contributors, however, may be tired as a result of the pressures of change, supporting and explaining changes that they may not have wanted; they may appreciate a rest. Do not forget to thank them.

Some staff may leave as a result of changes. Although it is unlikely (we would hope) that people will resign because the building is giving them a problem of some kind which cannot be rectified or tolerated, there may be building-related changes, such as working in greater or lesser proximity to someone else, that prove insurmountable. 'Exit interviews' should be arranged by HR staff with all who leave so that similar problems can be avoided in the future. Of course it would be better to be able to identify and deal with such issues before reaching this point but not everyone finds it easy to discuss these matters.

Hopefully the investment made in the building will result in staff turnover being reduced; indeed improved staff satisfaction and retention may be a major intended outcome of the improvements. It will be important to be able to compare staff turnover rates before, during and after major works.

When staff leave, they take with them 'corporate memory' so it is worth trying to retain as much of that before they depart. However, new staff will bring their own ideas and experiences and it will be important to try to capture and benefit from those. Enthusiastic staff are hard to attract so it is vital that such enthusiasm be encouraged, developed and maintained. Such people should be brought into discussions about the future – they will bring a fresh mind to matters.

New staff will also, of course, need and expect to receive induction and training in the ways things are done (or should be); this will provide opportunity to question why things are done the way they are and to suggest possible alternatives. Such suggestions should be sought routinely; they should be treated with respect and responded to after due consideration. An immediate response along the lines of 'We tried that and it didn't work' will serve to choke off further suggestions.

A climate in which people are encouraged to continually review how things are going, or have gone, will almost certainly give rise from time to time to proposals for organisational and other changes which may have building-related implications. A learning organisation will expect almost constant challenge. This may result in significant organisational and building-related churn. Churn is hard to define – it is a measure of the amount of change, whether it be in magnitude or frequency or direction. Change, and dealing with it, was discussed in Chapter 5.

Time and tide

Whatever is happening in, to and around a building, time is a constant consideration; it moves inexorably on.

> *Nae man can tether time or tide.* (Burns, 1791)

> *There is a tide in the affairs of men, Which, taken at the flood, leads on to fortune.* (Shakespeare, 1599)

Carpe diem -- 'seize the day'.

Our buildings will deteriorate over time; people and organisations will change and grow (and they will also decay and die). People will come and go. It is therefore important for the maintenance manager to keep abreast of developments and an ear to the ground. He or she will then be well placed to propose and implement changes to buildings, how they are configured and how they are maintained in the light of a changing context, without feeling at the mercy of events. The unexpected will still happen, and some of the expected will not happen, but there will be fewer surprises and they will be dealt with better.

It can be good to seize the moment, to support a good idea for the benefit of the individual seeing such a positive response. The value of 'quick wins' should not be overlooked in gaining goodwill and support for bigger and perhaps less attractive changes later.

The value of a celebratory party should also not be underestimated. Some people thrive on parties – they are good opportunities to socialise and to develop deeper understanding of colleagues than is often possible in the working environment. They also provide opportunities for those involved with projects to mark their completion, and of course to show continuing commitment and interest in further projects. They make for good marketing and publicity too – lots of happy smiling people drinking champagne, cutting ribbons, planting trees and so on. It may be appropriate therefore to think also of linking building-related projects to significant dates in an organisation's history. Silver, gold and diamond anniversaries are well worth marking, and centenaries especially. So it is worth thinking and planning ahead; these may provide good 'hooks' to help gain support for more major proposals, which will demonstrate very tangibly the organisation's faith in the future.

Checklist

This is the last checklist – but only in the sense of it being the last in this book. It may be the first of many that a maintenance manager may choose to use, or adapt, in the course of, and with the aim of, bringing buildings to, and keeping them in, a position where they meet users' needs.

1. What is the building supposed to be doing?
2. How well is it doing that?
3. What perhaps ought the building to be doing?
4. How well is it able to do that?
5. What do people want and/or need?

6. What plans are already in place; how well do they fit?

7. Who should be involved in taking matters forward?

8. Where are the priorities?

9. When should works be carried out?

10. How will it be known that matters have improved?

11. Why should we be doing this?

12. Who will benefit; by how much; and when?

Summary

This final chapter has reminded us that new and improved facilities need to be kept continually up to date and up to scratch. Techniques and timings have been considered and thoughts offered about motivation.

The book as a whole has examined a range of activities that might fall into the remit of a maintenance manager, building manager or facilities manager. It has considered matters through from the design of a new building or for the rehabilitation of an existing building, through the carrying out and supervision of the same, to its maintenance. Substantial space has been given to deliberating on what can go wrong, and why, and what to do about it. The enduring need to keep in mind ongoing, cyclical and new concerns has also been addressed, in terms of inspection, review and reporting. Particularities of dealing with matters in existing and occupied buildings have also been stressed. Throughout, there has been an emphasis on the human dimension, a reminder that buildings are for people.

It is hoped that the book will be seen not as the 'last word' in building maintenance but more as a starting point or stimulus for thought about priorities and how maintenance may best be achieved.

References

Argyris, C. & Schon, D.A. (1978) *A Theory of Action Perspective*. Addison-Wesley, Reading, MA.

Armstrong, M. (1994) *How to be an Even Better Manager* (4th edn). Kogan Page, London.

Bordass, B., Leaman, A. & Ruyssevelt, P. (2001) Assessing building performance in use 5: conclusions and implications. *Building Research and Information*, **29**(2), 144–157.

Bordass, W. & Leaman, A. (2005) Making feedback and post-occupancy evaluation routine: a portfolio of feedback techniques. *Building Research and Information*, **33**(4), 347–352.

Burns, R. (1791) Tam o'Shanter 1.57. In: *Oxford Dictionary of Quotations* (4th edn) (1992). Oxford Universtiy Press, Oxford, p. 163.

Buro Happold. *Post Occupancy Evaluation & Monitoring*. www.burohappold.com/BH/SRV_BLD_SC_postoccupancyevaluation.aspx. Accessed 26 February 2009.

Fewings, P. (2009) *Ethics for the Built Environment*. Taylor & Francis, Abingdon.

Handy, C. (1989) *The Age of Unreason*. Hutchinson Business, London.

Higher Education Funding Council for England (2006) *Guide to Post Occupancy Evaluation*. Available at www.smg.ac.uk/documents/POEBrochureFinal06.pdf. Accessed 26 February 2009.

Hoxley, M. & Rowsell, R. (2003) *The effective use of video in construction technology education*. CIB W89 International Conference on Building Education and Research (BEAR 2003), University of Salford, Salford, UK, April 2003, pp. 658–668.

Jaunzens, D., Grigg, P., Cohen, R., Watson, M. & Picton, E. (2003) *Building Performance Feedback: Getting Started,* BRE Digest 478. BRE, Watford.

Kolb, D.A. (1984) *Experiential Learning: Experience as the Source of Learning and Development.* Prentice-Hall, London.

Landsberger, H.A. (1958) *Hawthorne Revisited.* Cornell University Press, Ithaca, NY.

Leaman, A. & Bordass, W. (1995-1998) Post-occupancy review of buildings and their engineering. In *Building Services: Journal of the Chartered Institute of Building Services,* CIBSE, various dates from July 1995.

Leaman, A. & Bordass, B. (2000) Productivity in buildings: the 'killer' variables. In: Clements-Croome, D. (ed.) *Creating the Productive Workplace.* E. & F.N. Spon, London, pp. 167–191.

Mayo, E. (1933) *The Human Problems of an Industrial Civilisation.* MacMillan, New York.

Mayo, E. (1949) *Hawthorne and the Western Electric Company: the Social Problems of an Industrial Civilisation.* Routledge & Kegan Paul, London.

Mumford, A. (1989) *Management Development: Strategies for Action.* Institute of Personnel Management, London.

Oseland, N. (2007) *BCO Guide to Post Occupancy Evaluation.* British Council of Offices, London.

Preiser, W.F.E., Rabinowitz, H.Z. & White, E.T. (1988) *Post-Occupancy Evaluation.* Van Nostrand, New York.

Preiser, W. & Vischer, J. (2005) *Assessing Building Performance.* Elsevier, Oxford.

Revans, R. (1980) *Action Learning: New Techniques for Management.* Blond and Briggs, London.

Roethlisberger, F.J. & Dickson, W.J. (1939) *Management and the Worker.* Harvard University Press, Cambridge, MA.

Scottish Funding Council (2007) *Capital projects: post-occupancy evaluation guidance.* www.sfc.ac.uk/information/information_funding/POE_Guidance_Nov_07.pdf. Accessed 26 February 2009.

Senge, P. (1990) *The Fifth Discipline: The Art and Practice of the Learning Organisation.* Random House, London.

Shakespeare, W. (1599) Julius Caesar, act 4, scene 3. In: *Oxford Dictionary of Quotations* (4th edn) (1992). Oxford Universtiy Press, Oxford, p. 593.

Usable Building Trust for PROBE, Building Use Studies Occupancy Surveys. www.usablebuildings.co.uk. Accessed 26 February 2009.

Veitch, J.A. (2000) Creating high-quality workplaces using lighting. In: Clements-Croome, D. (ed.) *Creating the Productive Workplace.* E. & F.N. Spon, London, p. 206–222.

Index